FLORA OF TROPICAL EAST AFRICA

SAPINDACEAE

F.G. DAVIES & B. VERDCOURT*

Trees, shrubs or rarely herbs, sometimes climbing by tendrils. Indumentum commonly of simple hairs, sometimes in tufts in the axils of leaf-veins, rarely fasciculate; glands of several kinds occurring on both vegetative and flowering parts. Leaves alternate, sometimes crowded (palm-like) at the apex of the tree, variously simple, ternate, biternate, pinnate or bipinnate, in the latter two kinds most often without a terminal leaflet at maturity; stipules lacking or small. Inflorescences axillary or cauliflorous, thyrsoidal, paniculate or racemose. Flowers regular or slightly zygomorphic, unisexual except in some *Dodonaea* but with non-functional organs of the other sex usually present. Calyx of (3–)4–5(–7) free or partially united imbricate or valvate sepals. Petals absent or 4–5, often with a basal claw, usually with a simple or elaborated scale on the inner face. Disk conspicuous, simple, or rarely a double ring, or reduced to a pair of glands. Stamens 5–20(–74), often 8, but sometimes variable in number within a species, inserted in most cases inside the disk or occasionally on its surface; filaments free, terete or barely flattened, glabrous or hairy; anthers variously ovoid-sagittate to spherical, sometimes pilose, the connective rarely glandular, dehiscing introrsely by longitudinal slits; pollen grains tricolporate. Ovary 1–8-locular; ovules usually 1–2 (rarely several) per locule; style apical, except in *Allophylus* where semi-gynobasic, entire or 2–3-lobed. Fruit a capsule, sometimes lobed, or drupe, often composed of one matured mericarp with the aborted remaining carpels visible at its base. Seed usually with a hard black or brown testa, which in one case contains stomata, often with a conspicuous fleshy aril or sarcotesta, without endosperm.

About 147 genera and 2000 species, typically tropical and strongly represented in the American, African and Asian tropical areas; a number of species are cultivated for their edible fleshy fruits.

Radlkofer (E.P. IV, 165: 1–1539 (1931–4)) devoted a lifetime to the monographing of the family. Recently Leenhouts, in preparation for Flora Malesiana, has revised many genera and reappraised the tribes erected by Radlkofer. In 1976 (Muller & Leenh. in The evolutionary significance of the exine) he suggested that Radlkofer's more primitive tribes, the Sapindeae, Thouineae and Paullineae, are actually the most derived, and the Dodonaeae are not derived but a relict group. In effect this reverses the sequence of Radlkofer, and this is the order adopted here. Leenhouts (1976) further considers that the Aceraceae and Hippocastanaceae are no more than tribes of the Sapindaceae. [This is probably basically correct but these groups are of such horticultural importance I would not wish to tamper with their position – B.V.]

The flowering biology of the Sapindaceae in East Africa is very poorly known and field observations are urgently needed for all genera and species. There are two main classes of sexual arrangement: dioecy, generally constant for a genus (but *Allophylus* has only some dioecious species) and monoecy. In the latter the tree has mass-flowering of all its inflorescences simultaneously, each inflorescence starting with male flowers which drop off by abscission of the pedicel shortly after anthesis. The next flowers to open are female, which if

* F.G. Davies left Kew in 1986 but had completed the work for this family save for *Allophylus*; the account has been brought up to date by B. Verdcourt who has also provided the treatment of *Allophylus* based on notes prepared by R.M. Polhill over thirty years ago. *Bottegoa* Chiov. has been removed to Ptaeroxylaceae.

fertilised remain on the inflorescence, which then reverts to male flowers, sometimes in large numbers and continuing for a long season. However, this sequence may be imperfect through one or more stages being absent in any particular individual in any given season. The sex of the flower bud appears to be determined at least partly by environmental factors and at a late stage, and the sex of an immature inflorescence is impossible to determine in those genera where flowers are morphologically hermaphrodite – more than half those in East Africa. The term staminode is used here for a stamen which fails to dehisce. However, a small percentage of pollen in the staminodes of at least one species has been shown to be fertile.

The structure of the fleshy seed-appendage has been the subject of some disagreements. Van der Pijl (Acta. Bot. Neerl. 6: 608 (1957)) argues that the structure is not homologous with the aril in other families and should be called an 'arillode'. Corner (The seeds of Dicotyledons, 1976) disagrees and continues to use the term aril, while considering it sarcotestal in origin in some genera. The term aril will be used here regardless of ontogeny. Fruits of several species are quite unknown and should be looked for in the field.

The East African genera belong to the following subfamilies and tribes:

Subfamily *Dodonaeoideae* (*Dyssapindaceae* of Radlkofer)
- Tribe *Dodonaeeae* – 1, *Dodonaea*
- Tribe *Doratoxyleae* – 2, *Filicium*; 3, *Zanha*
- Tribe *Harpullieae* – 4, *Majidea*

Subfamily *Sapindoideae* (*Eusapindaceae* of Radlk.)
- Tribe *Cupanieae* – 5, *Aporrhiza*; 6, *Lychnodiscus*; 7, *Blighia*; 8, *Eriocoelum*; 9, *Haplocoelopsis*
- Tribe *Nephelieae* – 10, *Stadmania*; *Litchi*; *Nephelium*; 11, *Pappea*
- Tribe *Schleichereae* – 12, *Macphersonia*; 13, *Lecaniodiscus*; 14, *Haplocoelum*; 15, *Camptolepis*
- Tribe *Melicocceae* – *Tristiropsis*; *Melicoccus*
- Tribe *Lepisantheae* – 16, *Lepisanthes*; 17, *Glenniea*; 18, *Placodiscus*; 19, *Pancovia*; 20, *Chytranthus*
- Tribe *Sapindeae* – 21, *Deinbollia*; 22, *Sapindus*
- Tribe *Thouinieae* – 23, *Allophylus*
- Tribe *Paullinieae* – 24, *Cardiospermum*; 25, *Paullinia*

Several species have been or are cultivated in East Africa, including several well known fruits; all are included in the keys below. Cultivated species belonging to genera dealt within the text will be found there.

Litchi chinensis Sonn. subsp. *chinensis* (*Nephelium litchi* Cambess.) (T.T.C.L.: 559 (1949); U.O.P.Z.: 332 (1949)) the lychee (litchee) has been cultivated at Kenya, Kiambu District, Nairobi, Closeburn, *Gillett* 22754, Kilifi District, 34 km. N. of Mombasa, Vipingo, *Bally* 8854 and Tanzania, Lushoto District, Amani, *Greenway* 949 & *Magogo* 736 & *Ngonyani* 16 & *Furuya* 139 and Morogoro, *Greenway* & *Eggeling* 8601 and Mengo District, Entebbe (Dale, Introd. Trees Uganda: 48 (1953); also Jex-Blake (Gard. E. Afr., ed. 4: 302 (1957)) mentions that 'trees grow well and fruit abundantly near Nairobi'). Monoecious evergreen tree to 35 m. Leaflets in 2–4(–5) pairs, elliptic or ± obovate, (3–)8–11(–16) cm. long, 1.8–4 cm. wide, mostly acuminate at the apex, cuneate at the base, ± glaucous beneath. Inflorescences 15–30 cm. long with individual cymules 5–12-flowered; flowers yellow. Fruit bright red to purplish when ripe, subglobose, ± 3 cm. long, nearly smooth to densely covered with flat pyramidal acute tubercles up to 1 mm. high. Seeds polished blackish brown, very minutely rugulose, ellipsoid, with slight neck near hilum, ± 1.8–2.5 cm. long, 1.2–1.8 cm. wide, shining, partly or completely enveloped in a pinkish, bluish white or pale yellow juicy aril up to ± 5 mm. thick; hilum circular, 5 mm. diameter, straw-coloured. Morton, Fruits of Warm Climates: 250 (1987) deals with some of the better known cultivars; seedless varieties exist. A full account of the genus is given in Fl. Males., ser. 1, 11: 653–658 (1994).

Nephelium lappaceum L. var. *lappaceum* (T.T.C.L.: 559 (1949); U.O.P.Z.: 377, fig. on 378 (1949)) the rambutan has been grown for its edible fruit at Lushoto District, Amani, *Greenway* 746; 952; *Soleman* in *A.H.* 6184; and Mengo District, Entebbe (Dale, Introd. Trees Uganda: 52 (1953)). Tree 10.5–27 m.; twigs glabrescent to persistently hairy. Leaves 1-foliolate or with 1–7, mostly 5, leaflets; petiole 1.5–12 cm. long; leaflets relatively broad, obovate-oblong, 5–30 cm. long, 2.5–11 cm. wide, obtuse, rounded, truncate to slightly emarginate at the apex or sometimes apiculate to slightly acuminate or rarely distinctly acuminate, mostly widest above the middle, the sides very curved, glabrous above, hairy on midrib or ± all over beneath. Inflorescences axillary or terminal, lax; petals absent. Fruits very variable in colour, crimson, purple, maroon, greenish, yellowish or orange-yellow, ellipsoid to subglobose, (3.5–)6–8 cm. long, 3.5 cm. wide, covered with soft fleshy ± curved, strap-shaped or spine-like appendages 0.5–2 cm. long from tubercles; aril white, translucent, juicy; seed oblong, flattened, acute. Morton, Fruits of Warm Climates: 262–265 (1987) lists numerous varieties.

Melicoccus bijugatus Jacq. (*Melicocca bijuga* L., T.T.C.L.: 559 (1949); U.O.P.Z.: 349 (1949)) the marroncillo, honey berry or Spanish lime, a native of S. America has been grown at Lushoto District, Amani, *Greenway* 950 and Entebbe (Dale, Introd. Trees Uganda: 50 (1953)). Glabrous tree to 25 m. tall and bole to 1.7 m. diameter, with spreading branches and smooth grey bark. Leaves 10–16 cm. long; rhachis flattened above, not winged; leaflets in 2–3 pairs, the lower ovate, ± 5–8(–12) cm. long, 2–4(–6) cm. wide, the upper elliptic-lanceolate to elliptic, 8–11 cm. long, 2.5–5 cm. wide, acute; in young and vigorous leaves the petiole and rhachis may be ± broadly winged and leaflets up to 18 cm. long, 8 cm. wide. Male inflorescences with several elongate racemose branches, 6–10 cm. long; female usually ± simple. Flower whitish, 5–8 mm. wide. Fruits green or yellow, sub-globose, plum-like, 2.5 cm. long, 1.5–2.5 cm. wide, with rather scanty flesh; seed yellowish white, large or 2 hemispherical seeds; aril salmon-coloured or yellowish.

Tristiropsis acuteangula Radlk. a native of Malesia to Queensland, Solomon Is., Guam and Palau Is. has been grown in Zanzibar, Dunga, Feb. 1929, *Greenway* in *A.D.* 27 (EA, K). It is not known who introduced it nor why – it has been grown for timber in the east. A tree 7.5–35(–53) m. tall, often buttressed; branchlets fulvous tomentose at first. Leaves bipinnate or upper ones at least forked, up to 2 m. long; petiole to 20 cm. long. Leaflets ovate to elliptic, 5–18 cm. long, 1.5–8 cm. wide, emarginate, obtuse or acute and often mucronulate at the apex, acute to obtuse at the base. Inflorescence up to 40 cm. long, densely adpressed brownish pubescent; peduncle to 15 cm. long; cyme-stalks up to 8 mm. long; cymules containing both male and female flowers; pedicels 3–10 mm. long. Flowers sweet-scented; petals cream, 2.5–3.5 mm. long, 2.2–2.5 mm. wide. Fruits yellowish-green to dark yellow, ellipsoid to subglobular, 2–3 cm. long, 1.4–2.5 cm. wide, widest at or above the middle, narrowed to shortly stipitate at the base, 3–4-angular or 3–4-ribbed in cross section, densely minutely ferruginous tomentose, not dehiscent.

Key to flowering specimens

Note. This and the key to fruiting specimens (on p. 5) apply only to species occurring in the Flora area. It should be noted that immature flowers are not directly comparable with mature ones: the petal-claw and scales only develop at a late stage. Similarly the aril, if present, develops late in the ripening fruit. The gross morphology of the fruit, if not the precise shape, can be determined in a young female flower by a cross-section of the ovary in which the number of locules and ovules is clearly visible.

Flowers appearing when plant is leafless:
- Disk cushion-shaped; stamens inserted into pockets in the disk; ovary 3-locular, locules 1-ovulate ··· 14. **Haplocoelum**
- Disk flat or cup-shaped; stamens inserted within the disk; ovary 2-locular, locules 2-ovulate ··· 3. **Zanha**

Flowers and leaves appearing together:
- Leaves simple:
  - Leaves sticky-glandular and shiny at least when young; margin entire ··· 1. **Dodonaea**
  - Leaves not sticky; margin entire, crenate or serrate:
    - Leaf-apex narrowly acuminate or acute, rarely bluntly acuminate; buds covered by imbricate sepals; leaf nervation not forked towards margin ··· 23. **Allophylus**
    - Leaf-apex obtuse, emarginate or rarely shortly acuminate; buds opening precociously; leaf nervation prominent and forked near margin ··· 11. **Pappea**
- Leaves compound:
  - Plant climbing by tendrils:
    - Leaves pinnate, the rhachis winged ··· 25. **Paullinia**
    - Leaves biternate, the rhachis unwinged ··· 24. **Cardiospermum**
  - Plant not climbing by tendrils:
    - Leaves bipinnate:
      - Petals 5; stamens inserted inside the disk; leaf-rhachis pubescent ··· 12. **Macphersonia**

Petals 4; stamens inserted outside the disk; leaf-rhachis with resin dots · · · · · · · · · · · · · · · · · **Bottegoa***
Leaves pinnate:
Stamens 4–5:
Petals absent in Flora area:
Lower leaflets slightly smaller than the upper:
Stamens inserted into pockets in the cushion-shaped disk; ovary 3-locular, locules 1-ovulate · 14. **Haplocoelum**
Stamens inserted inside the disk; ovary 2-locular, locules 2-ovulate · · · · · · · · · · · 3. **Zanha**
Lower leaflets tending to resemble stipules, much reduced · 17. **Glenniea**
Petals 5 (sometimes minute):
Leaf-rhachis winged (in Flora area); petals minute, without a scale · · · · · · · · · · · · · · 2. **Filicium**
Leaf-rhachis unwinged; petals with hair-fringed scales · 16. **Lepisanthes**
Stamens 6–14(–30):
Petals 4, flowers radially or bilaterally symmetrical:
Leaves trifoliolate · 23. **Allophylus**
Leaves usually with 2 or more pairs of leaflets, rarely abnormally trifoliolate:
Flowers (calyx and curvature of styles and stamens) bilaterally symmetrical:
Leaflets in 1–5 pairs; inflorescences borne on slender branches; calyx campanulate; usually branched trees or shrubs; petal-scale lobed or laciniate · · · · · · · · · · · 19. **Pancovia**
Leaflets in 3–12 pairs; inflorescences borne on old wood often near to the ground; calyx urceolate; usually unbranched palm-like trees or shrubs with crown of leaves; petal-scale entire or 2-lobed · · 20. **Chytranthus**
Flowers radially symmetrical:
Inflorescence-axis glabrous; cultivated plant · **Melicoccus**
Inflorescence-axis fasciculate-hairy; native plant · 4. **Majidea**
Petals absent or 5; flowers symmetrical; pubescence of unbranched hairs or absent:
Petals absent:
Sepals opening precociously in bud before the elongation of the filaments:
Leaflet-apices rounded or emarginate; inflorescence a narrow thyrse · · · · · · 10. **Stadmania**
Leaflet-apices acuminate; inflorescence a broad panicle · · · · · · · · · · · · · · · · · **Litchi**
Sepals remaining imbricate around elongated and folded filaments in bud:
Cultivated tree; ovary warty (obviously so just after fertilisation) · · · · · · · · · · · **Nephelium**

* Now placed in Ptaeroxylaceae but retained in key since it could well be looked for in Sapindaceae

Native trees; ovary not warty:
  Inflorescence arising from the trunk or branches; fine vein-reticulum of leaflets very prominent; ovary deeply 3-lobed · · · · · · · · · · · · · 18. **Placodiscus**
  Inflorescence from leaf-axils; fine vein-reticulum not noticeably prominent; ovary not lobed · · · · · · · · · · · · 13. **Lecaniodiscus**

Petals 5:
 Petals 2–5 × as long as sepals, with short free broad hairy scales; stamens 8; leaves with lowermost pair stipule-like (U 1) · · · · · 8. **Eriocoelum**
 Without above characters combined:
  Filaments glabrous throughout:
   Lowest leaflets very reduced, stipule-like; ovary 2-lobed; inflorescence racemose, axillary · 9. **Haplocoelopsis**
   Lowest leaflets usually smaller than the upper but not stipule-like; ovary 3-lobed or if 2-lobed the inflorescence a large panicle over 20 cm. long:
    Leaflets toothed or undulate; inflorescence a large panicle 14–50 cm. long; disk unlobed, of 2 concentric rings · · · · · · · · · · · · 6. **Lychnodiscus**
    Leaflets entire; inflorescence a fascicle of cymes up to 6 cm. long; disk lobed, single · · · · · · · · · · · · · · · 15. **Camptolepis**
  Filaments hairy at least for part of their length:
   Buds opening precociously to reveal sessile anthers; leaflets 1–5 pairs · · 7. **Blighia**
   Buds covered by imbricate sepals; leaflets 3–18 pairs:
    Tree with palm-like habit; stamens 14–20 · · · · · · · · · · · · · · · · · · · 21. **Deinbollia**
    Branched trees; stamens 6–10:
     Ovary 2-lobed; filaments folded or curved in bud; native plants:
      Vegetative parts and sepals greyish-green or rusty-tomentose · · · 5. **Aporrhiza**
      Vegetative parts and sepals glabrous 16. **Lepisanthes**
     Ovary 3-lobed, filaments undeveloped in bud; cultivated plants:
      Petals without scales · · · · · · · · · 22. **Sapindus**
      Petals with 2 substantial scales · · **Tristiropsis**

KEY TO FRUITING SPECIMENS

Lowest leaflets stipule-like:
 Fruits loculicidally dehiscent with thick woody valves · · · 8. **Eriocoelum**
 Fruits baccate or of mericarps:
  Fruit glabrous, globose, indehiscent · · · · · · · · · · · · · · 9. **Haplocoelopsis**
  Fruit velvety, of 1–2 dehiscent mericarps · · · · · · · · · · · 17. **Glenniea**
Lowest leaflets not stipule-like, or leaves simple:

Fruit an inflated papery capsule or with papery wing or wings; leaves not simply pinnate:
Fruit an inflated capsule; climber with tendrils · · · · · · 24. **Cardiospermum**
Fruit winged; erect plants:
Fruit discoid with spongy centre and encircling wing; leaves bipinnate · **Bottegoa***
Fruit 1–3-lobed, each lobe with a wing; leaves simple 1. **Dodonaea**
Fruit not papery; leaves various:
Plant climbing by tendrils; fruit a 3-valved, red, obovoid capsule · 25. **Paullinia**
Plant without tendrils; fruit without this combination of characters:
Fruit dehiscent with 2–3 valves and 2–3 seeds:
Seeds without aril, blue-black velvety; hairs on the inflorescence fasciculate · · · · · · · · · · · · · · · · 4. **Majidea**
Seeds with a fleshy orange or yellow aril, glabrous; hairs simple:
Capsule externally glabrous, reddish, drying black 7. **Blighia**
Capsule externally pubescent, brownish:
Fruit of 2 compressed mericarps; leaflets entire 5. **Aporrhiza**
Fruit a 3-lobed capsule; leaflets undulate or crenate or toothed · · · · · · · · · · · · · · · · · · 6. **Lychnodiscus**
Fruit either indehiscent or if dehiscent then containing 1 locule and 1 seed at maturity:
Fruit subglobose, 1.2–3 cm. wide, 3-locular with an apical areole around the style remnant; seeds with a fleshy aril; hilum placed at base of seed on adaxial side and ± vertical · · · · · · · · · · · · 15. **Camptolepis**
Fruit and seeds not as above:
Fruit 3–5-angled or lobed, composed of 3–5 carpels; seeds without arils:
Fruit obovoid, 3-angled, greenish, glabrous (cultivated) · **Tristiropsis**
Fruit ovoid or subglobose with rounded lobes to star-shaped, variously coloured, pubescent (native):**
Leaflets with exceptionally fine network and of prominent veins (**T** 3, E. Usambara Mts.; **T** 6, Kilombero) · · · · · · · · · · · · · · · · · · 18. **Placodiscus**
Leaflets without such a prominent network of veins:
Leaflets in 1–5 pairs; inflorescences borne on slender branches; usually branched trees or shrubs · · · · · · · · · 19. **Pancovia**
Leaflets in 3–12 pairs; inflorescences borne on older wood, often near the ground; usually unbranched, palm-like trees or shrubs or single-stemmed subshrubs; leaves in apical crown · · · · · · · · · · · · 20. **Chytranthus**
Fruit either deeply divided into 1–3 mericarps of which 1–2 often remain undeveloped at the base of the fertile one, or a single drupe derived from the whole ovary by internal reduction to one seed; seed with or without aril:

* See footnote on p. 4.
** Ripe fruit unknown for *Placodiscus* but judging by nearest congeners, taxa would key out here.

Fruit warty or spiny outside with a single glabrous seed enclosed by a discrete succulent aril (cultivated):

Fruit warty; leaflets shiny above, mat and dull beneath · **Litchi**

Fruit spiny; leaflets similar on both surfaces **Nephelium**

Fruit neither warty nor spiny:

Fruit of 1–3 fertile mericarps, usually 1–2 abortive and small remaining at the base of the fertile one(s):

Leaves simple or trifoliolate; mericarps globose or ovoid, 3–10 mm. wide:

Leaves always simple, usually obtuse or emarginate at the apex, pubescent beneath, with prominent close obviously branched nerves; fruits velvety tomentose, ± 1 cm. wide · · · 11. **Pappea**

Leaves trifoliolate or in 4 species simple but then ± glabrous beneath (save in a var. of *A. congolanus*); nerves less prominent and less often branched; fruits glabrous or sparsely pubescent, not velvety, smaller · · · · · · · · · · · · 23. **Allophylus**

Leaves pinnate; fruits mostly larger:

Fruits pubescent or velvety tomentose:

Fruits in extensive branched inflorescences 7–60 cm. long:

Fruit brown or red-brown; habit usually palm-like with terminal clusters of leaves and inflorescences; leaves up to 12-foliolate; wild · · · · · · · 21. **Deinbollia**

Fruit dark, blackish; habit not as above; leaves 4–6(–8)-foliolate; cultivated 22. **Sapindus**

Fruits in unbranched inflorescences up to 20 cm. long or in very narrow thyrses with side branches very reduced and ± nodular (branched trees or shrubs):

Fruit with sparser indumentum leaving the actual surface visible; aril absent · · · · · · · · · · · · · · · · 19. **Pancovia**

Fruit with velvety tomentose covering the surface; seeds covered by red (in life) aril · · · · · · · · · · · · · · 10. **Stadmania**

Fruit glabrous:

Leaflets in 1–2(–3) pairs; fruit crimson; native · 16. **Lepisanthes**

Leaflets in 3–6 pairs:

Fruits yellowish; native · · · · · · · · · 21. **Deinbollia**

Fruit blue, black or brown; cultivated 22. **Sapindus**

Fruit a single 1-seeded capsule or drupe, indehiscent or irregularly dehiscent; without abortive mericarps at the base:

Leaves trifoliolate or in a few species simple (included here because the abortive mericarps can be difficult to see, but the gynobasic style is often persistent at the base of the remaining ones) · · · · 23. **Allophylus**
Leaves with 4 or more leaflets:
 Fruit plum-like, ± 2.5 cm. long; plant cultivated · · · · · · · · · · · · · · · · · · · **Melicoccus**
 Fruit usually smaller; native plants:
  Leaves bipinnate · · · · · · · · · · · · · · · 12. **Macphersonia**
  Leaves pinnate:
   Seeds without aril:
    Leaf-rhachis winged; leaflets lanceolate to linear · · · · · · · 2. **Filicium**
    Leaf-rhachis unwinged; leaflets ovate to elliptic · · · · · · · · · · 3. **Zanha**
   Seeds with fleshy aril (inconspicuous when dry):
    Fruit glabrous; leaflets noticeably asymmetric at their base · · · · 14. **Haplocoelum**
    Fruit pubescent; leaflets not noticeably or only slightly asymmetric · · · · · · · · · · · · · 13. **Lecaniodiscus**

1. DODONAEA

Mill., Gard. Dict., ed. 4, 1, unpaginated (Dodonaea) (1754); Radlk. in E.P. IV, 165: 1350–1404 (1933); Leenh. in Blumea 28: 271–289 (1983); J.G. West in Brunonia 7: 1–194 (1984)

Small evergreen trees or shrubs, sometimes ericoid, with simple or compound leaves, sometimes with conspicuous glandular varnish. Flowers regular, bisexual or unisexual, individual plants bearing one or more types of flower; sepals 3–4(–7), valvate or imbricate, free (in Africa) or coherent at the base, deciduous or persistent; petals absent. Stamens 5–8(–10), inserted into the outer surface of the circular disk; filaments glabrous, much shorter than the anthers; anthers oblong, sagittate at the base. Ovary rounded or lobed, 2–4(–6)-locular, each locule with 2 ovules; style as long as to much longer than the ovary; ovary rudiment in the male flower minute. Fruit a papery capsule, often winged, dehiscent by splitting along 2–3 central septa. Seeds 1(–2) per valve, black, globose or lenticular, without an aril.

A genus of 50 species in Australia, one in New Guinea and Queensland, one throughout the tropics and one endemic to Madagascar.*

D. viscosa *Jacq.*, Enum. Syst. Pl.: 19 (1760); Radlk. in E.P. IV, 165: 1363 (1933); T.T.C.L.: 557 (1949); U.O.P.Z.: 235 (1949); Jex-Blake, Gard. E. Afr., ed. 4: 111, 356 (1957); F.W.T.A., ed. 2, 1: 724 (1958); Exell in F.Z. 2: 542 (1966); Vollesen in Opera Bot. 59: 58 (1980); Leenh. in Blumea 28: 285, fig. 1a (1983); J.G. West in Brunonia 7: 30, fig. 22a (1984) & in Fl. Austr. 25: 120, fig. 28a (1985); Leenh. in Fl. Males., ser. 1, 11: 526 (1994); Beentje, K.T.S.L.: 417, map (1994); F. Friedmann in Fl. Masc. 76: 3, t. 1/1–3 (1997).** Lectotype, chosen by J.G. West: Jamaica, Old Harbour; Sloane, Voy. Is. Madera, Barbados, Nieves, S. Christ., Jamaica 2: 27, t. 162/3 (lecto.!); based on Sloane Herb. 5: 97 (BM, typotype)

* I have deviated considerably from Dr. Davies's treatment; she considered there were two species in East Africa one coastal and the other mainly inland but after much thought I have treated these taxa as varieties. See note at end of species. [B.V.]
** For references which actually refer only to var. *angustifolia* see p. 11.

Monoecious or dioecious semi-prostrate shrub or small tree, 0.5–9 m. tall; bark black or brown, ± rough; twigs black or reddish brown, glandular, glabrescent, ridged and developing vertical fissures. Leaves simple, variable; blades oblanceolate or broadly to narrowly elliptic, (1–)4–13 cm. long, (0.4–)1.3–4.2 cm. wide, widest at or above the midpoint, obtuse and minutely apiculate at apex, gradually narrowed below the middle to a narrowly cuneate base; margin entire, recurved (in dried specimens); both surfaces glabrous and glandular, coated especially when young with viscid glandular exudate; lateral nerves 15–20(–30), often indistinct; petioles 0–2.5 mm. long. Flowers greenish yellow, unisexual or bisexual, in loose thyrsoid panicles at the ends of twigs; pedicels 0.8–1.5 cm. long in flower and fruit, glabrous and glandular. Male flowers: sepals 3–4, straw-coloured or greenish yellow, 2–2.5 mm. long, spreading; stamens 7(–9); anthers 2–3 mm. long; ovary rudiment minute. Female flowers: calyx green, cup-like but sepals free; staminodes lacking or with anthers ± 1.5 mm. long; ovary green, oblong in outline, flattened, bilobed, glandular; style 2–3-lobed, equalling the ovary; disk inconspicuous. Bisexual flowers similar: sepals 3–4, greenish yellow, ovate, 2–2.5 mm. long, spreading; anthers 1.6–1.9 mm. long. Fruit disk-shaped, the body 1.5–2.3 cm. long, 8–11 mm. wide, with an encircling papery venose wing 3–7 mm. wide or a few with 3(–4) locules and wings. Seeds 2 per locule, black, subspherical, or somewhat compressed with a raised dorsal rim, 2–3 mm. diameter, 1.5–3 mm. thick.

var. **viscosa**

Small tree or shrub 1.5–4 m. tall. Leaves oblanceolate or narrowly elliptic, 6–12.5 cm. long, 1.8–4.2 cm. wide, mostly widest above the middle and relatively broader. Flowers whitish, mostly bisexual; scar of fallen sepals beneath fruit strongly bilobed. Fruit white or brown to straw-coloured or greenish white, sometimes yellow or reddish, mostly 2-winged, 1.5–2.3 cm. long, 1.8–2.5 cm. wide. Seeds subglobose, 3 mm. diameter, 2–3 mm. thick, mostly rolling easily when dropped on a flat surface.

KENYA. Mombasa District: Nyali Beach, 8 Aug. 1965, *Strange* 74!; Kilifi District: Fundi Isa, *R.M. Graham*!; Lamu District: Lamu I., Lamu to Shela, 24 Aug. 1973, *Gillett* 20317!

TANZANIA. Tanga, near seashore, 4 Nov. 1929, *Greenway* 1828!; Pangani, Sept. 1955, *Mgaza* 85!; Rufiji District: Mafia I., Kanga seashore, 11 Aug. 1937, *Greenway* 5052!; Zanzibar, Chwaka beach, 6 Nov. 1959, *Faulkner* 2392!; Pemba I., Mkoani, 7 Sept. 1929, *Vaughan* 638!

DISTR. **K** 7; **T** 3, 6, 8; **Z**; **P**; West African coasts of Senegal, Ghana and Nigeria, East Africa as far south as Mozambique and Madagascar, also India, Malesia, Australia, Pacific Is. and America

HAB. Coastal bushland, on sand dunes and coral rock mostly just above high water mark; but also *Casuarina* woodland, *Grewia glandulosa* scrub and plantations; 0–30(–75) m.

SYN. *Dodonaea viscosa* Jacq. subsp. *viscosa*; J.G. West in Brunonia 7: 37, fig. 22A (1984) & in Fl. Austr. 25: 121, fig. 28A (1985)

var. **angustifolia** (*L.f.*) *Benth.*, Fl. Austral. 1: 476 (1863); Radlk. in E.P. IV, 165: 1368 (1933); C.D. Adams, Fl. Pl. Jamaica: 439 (1972). Lectotype, chosen by J.G. West: Smith Herb. 673.3 (LINN, lecto.)

Tree or shrub 0.5–7.5(–9) m. tall, usually much-branched with black bark. Leaves broadly to narrowly elliptic or linear-oblong to narrowly oblanceolate, 1.5–10(–13) cm. long, 0.6–2.7(–3.5) cm. wide, narrowed gradually and rounded to an acute and weakly apiculate tip. Flowers greenish yellow, mostly unisexual; scar of fallen sepals beneath fruit usually annular or slightly lobed. Fruit mostly red-tinged or mottled or crimson to purplish but can be white, brown or reddish green, predominantly 2-winged but often with 3 (or even 4) wings. Seeds compressed globose, ± 3 mm. diameter, 1.5–1.8 mm. thick, with obtuse dorsal keel, not easily rolling when dropped on a flat surface. Fig. 1.

UGANDA. Ankole District: Ruizi R., 15 Nov. 1950, *Jarrett* 211!; Kigezi District: Nyakabande, 19 Dec. 1933, *A.S. Thomas* 1083!; Mbale District: Elgon, Kaburoron [Kaburon], Jan. 1936, *Eggeling* 2481!

FIG. 1. *DODONAEA VISCOSA* var. *ANGUSTIFOLIA* — **1**, flowering branch, × $^{2}/_{3}$; **2**, male flower, × 6; **3**, female flower, × 6; **4**, vertical section of female flower, × 6; **5**, infructescence, × $^{2}/_{3}$; **6**, seed, × 4. 1, 2, from *Chase* 2182; 3, 4, from *Chase* 1671; 5, 6, from *Swynnerton*. Reproduced from F.Z. 2, t. 117 (1966).

KENYA. Northern Frontier Province: Uaraguess, 3 Dec. 1958, *Newbould* 3120!; Turkana District: Kachagalau [Kacagalau] Forest Reserve, 15 Jan. 1959, *Lang Brown* 102!; N. Nyeri District: Nanyuki, 2 July 1956, *Verdcourt* 1514!

TANZANIA. Arusha District: Mt. Meru, E. slope of Tulusia [Tululusia] Hill, 12 Mar. 1968, *Greenway & Kanuri* 13372!; Kondoa District: S. slope of Kinyassi Mt., 8 Feb. 1928, *B.D. Burtt* 1358!; Songea District: Matengo Hills, 8 km. N. of Miyau, 22 May 1956, *Milne-Redhead & Taylor* 8924A; Zanzibar, Muyuni, 19 Oct. 1930, *Vaughan* 1645!

DISTR. **U** 1–4; **K** 1–7; **T** 1–8; **Z**; tropical Africa from Zaire in the west (cultivated only in Ghana, Nigeria and Cameroon) to Ethiopia and Somalia in the east, widespread in south tropical Africa and South Africa, also Madagascar and all parts of the tropical and subtropical world

HAB. Grassland, bushland, woodland and thicket but also evergreen forest, clearings in *Juniperus* forest etc., woodland or forest on lava, often in rocky places, stony hillsides, persisting in disturbed areas, even bad wasteland; (0–)500–2700 m.

SYN. *Ptelea viscosa* L., Sp. Pl.: 118 (1753)*. Type: Sri Lanka, *Herb. Hermann* (BM!, lecto.)

Dodonaea viscosa (L.) L., Mant. Pl. 2: 149, 228 (1771) excl. cit. Jacq., *nom. illegit.*, *non* Jacq.

D. angustifolia L.f., Suppl. Pl.: 218 (1782); Leenh. in Blumea 28: 280 (1983); Blundell, Wild Fl. E. Afr.: 136, t. 472 (1987); Noad & Birnie, Trees of Kenya: 263, fig. on 264 (1989); Vollesen in Fl. Ethiopia 3: 491, fig. 125.2 (1990); Leenh. in Fl. Males., ser. 1, 11: 523, fig. 24c (1994) (full description and synonomy); Beentje, K.T.S.L.: 416, fig. & map on 417 (1994)

[*D. viscosa* sensu Baker in F.T.A. 1: 433 (1868); Radlk. in E.P. IV, 165: 1363 (1933) pro parte; F.P.N.A. 1: 523, t. 51 (1948); T.T.C.L.: 557 (1949) pro parte; F.P.S. 2: 339, fig. 126 (1952); I.T.U., ed. 2: 378, t. 19 (1952); C.F.A. 2: 91 (1954); F.W.T.A., ed. 2, 1: 724 (1958) pro parte; Hauman in F.C.B. 9: 382, t. 39 (1960); K.T.S.: 511, t. 30 (1961) pro parte; F.F.N.R.: 224 (1962); Exell in F.Z. 2: 542, t. 117 (1966) pro parte; Fouilloy & Hallé in Fl. Cameroun 16: 187 (1973) & in Fl. Gabon 23: 187 (1973); Vollesen in Opera Bot. 59: 58 (1980); Troupin, Fl. Rwanda 2: 304, fig. 95/1 (1983), *non* Jacq. sensu stricto]

D. viscosa Jacq. subsp. *angustifolia* (L.f.) J.G. West in Brunonia 7: 39 (1984) & in Fl. Austr. 25: 122 (1985)

NOTE. I at first accepted the separation into two species proposed by Leenhouts (who can hardly be called a 'splitter') but found the calyx scar character he uses does not always work nor does the character of the seed rolling emphasised by Davies; it is particularly difficult to use where both taxa occur, e.g. in Zanzibar or the Mozambique coast. Subtle leaf characters do emerge after examining much material but are variable. I have not felt able to retain two species. The idea of ecological subspecies has also been rejected although the separation of coastal and upland populations is certainly valid to a large extent; but it involves the supposition that both subspecies have had a different geographical history and that *angustifolia* in Jamaica and East Africa say are more closely related to *angustifolia* in Australia and India than they are to the lowland populations in the same territories. I prefer to maintain a wide concept of *D. viscosa* with a tendency to produce varieties, mainly two in much of the world but more elsewhere (there are distinctive ones in Australia and South Africa). The mechanism of the variation could be simple but involve some ecological selection. The species would be easy to grow with a view to genetic research. It is possible that all the other subspecies of *D. viscosa* are genuine subspecies in which case the two varieties treated above belong to subsp. *viscosa*. [B.V.]

The wood is hard and used in some parts of the world for walking sticks etc.; it is used for hedging in Kenya and the leaves have been used medicinally but reported stock deaths seem rather unlikely although the leaves may contain an alkaloid (see Watt & Breyer-Brandwijk, Med. Pois. Pl. S. & E. Afr., ed. 2: 931 (1962)).

* The references cited by Linnaeus cover both taxa but the cultivated Hort. Cliff. specimen (Hort. Cliff. *Dodonaea* 1 (BM)) is probably var. *angustifolia* and the Hermann specimen definitely is hence its choice as lectotype. In his original account (Fl. Zeyl.: 58 (1747)) Linnaeus used the generic name *Dodonaea*. West includes *Ptelea viscosa* L. under subsp. *burmanniana* (DC.) J.G. West but Leenhouts includes *D. burmanniana* DC. under *D. angustifolia*.

2. FILICIUM

Hook.f. in G.P. 1: 325 (1862); Radlk. in E.P. IV, 165: 1426 (1933)

Pteridophyllum Thwaites in Hook., Journ. Bot. 6: 65, t. 1B (1854), *nom. illegit.*, *non* Sieb. & Zucc. (1843)

Elegant trees; twigs glabrous, dark grey, warty. Leaves simple or 1–many-times pinnate, usually including a terminal leaflet; petiole and rhachis winged or not; leaflets with wax dots, glossy. Inflorescences axillary, narrowly paniculate, each containing both male and female flowers. Sepals 5, free; petals minute, without claw or scales. Disk circular, densely shaggy-white-hairy. Stamens 5; anthers dorsifixed, sagittate. Ovary globose; ovary-rudiment in male flower absent. Fruit a 1-seeded drupe, purple, glabrous and shining. Seed without aril; cotyledons contorted.

A genus of 3 species, 2 of which are endemic to Madagascar and neighbouring islands. The third occurs from India and Sri Lanka to East Africa, where it is certainly native despite its also being cultivated here and in the Pacific Islands.

F. decipiens (*Wight & Arn.*) *Thwaites*, Enum. Pl. Zeyl.: 408 (1864); T.T.C.L.: 557 (1949); Dale, Introd. Trees Uganda: 42 (1953); Jex-Blake, Gard. E. Afr., ed. 4: 224 (1957); K.T.S.: 511 (1961); Exell in F.Z. 2: 534, t. 114 (1966); Capuron in Mém. Mus. Nat. Hist. Nat., Paris, 19: 38, t. 7/1–13 (1969); Noad & Birnie, Trees of Kenya: 265, fig. on 266 (1989); Vollesen in Fl. Ethiopia 3: 493, fig. 125.4 (1990); Leenh. in Fl. Males., ser. 1, 11: 754, fig. 86 (1994); Beentje, K.T.S.L.: 417, fig. & map (1994). Types: India, *Wight** 520 & 521 (K!, syn.) & Sri Lanka; Burm., Thes. Zeyl.: 99, t. 45 (1737) (syn.!)

Slender tree 4.5–20(–30) m. tall; bark smooth, whitish. Petiole winged, 1.5–11 cm. long; rhachis 15–30 cm. long, winged each side to 1 cm. wide, tapering below each leaflet pair; leaflets in 5–10 pairs, sessile, lanceolate to linear or narrowly oblong-elliptic, 5–16 cm. long, 0.5–4 cm. wide, acute, entire but crisped-undulate at the margin, glabrous, resinous when young; lateral nerves in 30–50 pairs, faint. Inflorescence 6–20 cm. long; pedicels 2–4 mm. long. Sepals 1.3 mm. long, herbaceous, puberulous inside, ciliate; petals yellow, 1.3 mm. long, ciliate, apparently caducous. Stamens exserted from the calyx; staminodes not exserted. Ovary 2-locular with 2 ovules, one only developing. Fruit ellipsoid or subglobose, 9–10 mm. long, 6–8 mm. wide, with persistent style and reflexed sepals. Fig. 2.

KENYA. Fort Hall District: Thika, Chania Gorge, 12 Apr. 1964, *Verdcourt* 4038B!; Kiambu District: Nairobi Falls, 17 July 1966, *Gillett* 17345!; Kiambu/Machakos District: Fourteen Falls, 6 Sept. 1952, *Verdcourt* 720!

TANZANIA. Lushoto District: Makuyuni, 16 Dec. 1935, *Koritschoner* 1441!; Ulanga District: Mahenge, Tabora, 3 Feb. 1932, *Schlieben* 1726!; Morogoro District: Kimboza Forest Reserve, 10 Apr. 1957, *Semsei* 2669!

DISTR.** **K** 1, 4, 7; **T** 2, 3, 6, 7; Ethiopia, Malawi, Mozambique, Zimbabwe, Madagascar, Comoro Is., India and Sri Lanka; also cultivated at Nairobi (arboretum, 22 Mar. 1963, *Karanja* 7) & Amani (Boma Garden, 26 Aug. 1929, *Greenway* 1689), Java, Sabah, Fiji, Samoa and Hawaii

HAB. Mixed forest in river valleys; 450–1600 m.

SYN. *Rhus decipiens* Wight & Arn., Prodr. Fl. Pen. Ind. Or. 1: 172 (1834)
Pteridophyllum decipiens (Wight & Arn.) Thwaites in Hook., Journ. Bot. 6: 66, t. 1B (1854)
Filicium elongatum Radlk. in P.O.A. C: 252 (1895); T.T.C.L.: 557 (1949). Type: Tanzania, Moshi District, Kilimanjaro, above Marangu, *Volkens* 1715 (B, holo.)

* Wight has two varieties α and β; *Wight* 521 is cited under the latter and presumably should be excluded from a lectotypification. [B.V.]

** I have not seen material from **K** 1, 7 or **T** 7. [B.V.]

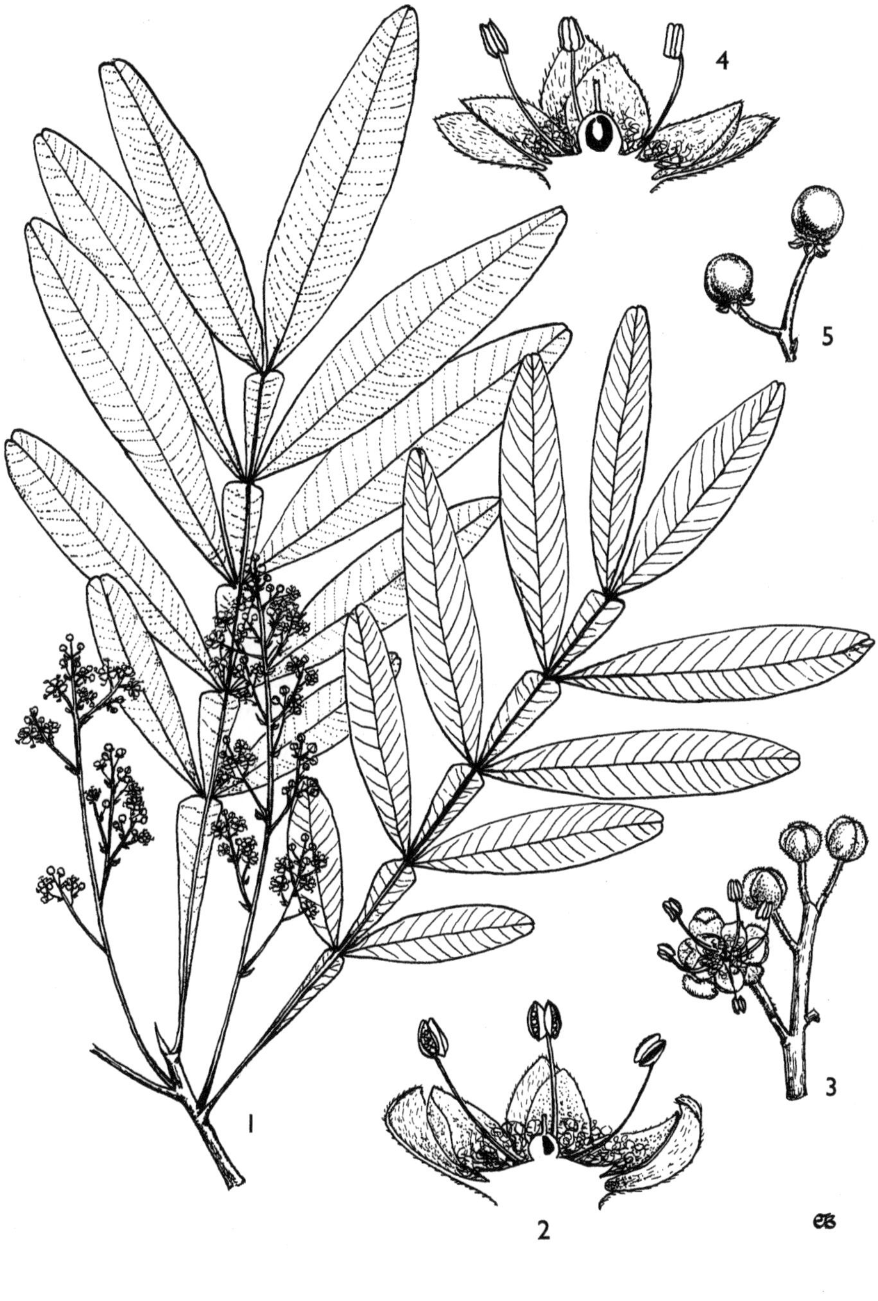

FIG. 2. *FILICIUM DECIPIENS* — **1**, flowering branchlet, × $^2/_3$; **2**, vertical section of male flower, × 16; **3**, male flower and flower-buds, × 5; **4**, vertical section of female flower, × 5; **5**, fruits, × 1. 1, from *Farquhar & Finn* in *G.H.S.* 26611; 2, 3, from *Wild* 5275; 4, from *Farquhar* 10/49; 5, from *Goldsmith* in *G.H.S.* 68162. Reproduced from F.Z. 2, t. 114 (1966).

NOTE. Male and female flowers in an inflorescence appear to be separated in time by a very small margin compared with other Sapindaceae and may even overlap. It would be of interest to have this confirmed (or not) in the field.

Capuron (Mém. Mus. Nat. Hist. Nat., Paris, 19: 40 (1969)) describes a forma *apterum* from Madagascar with unwinged petioles and leaf-rhachises and Leenhouts figures such a plant.

3. ZANHA

Hiern, Cat. Afr. Pl. Welw. 1: 128 (1896); Radlk. in E.P. IV, 165: 1420 (1933)

Dialiopsis Radlk. in E. & P. Pf., Nachtr. 3: 207 (1907) & in E.P. IV, 165: 1419 (1933)

Deciduous dioecious trees or sometimes shrubs, flowering generally before the leaves emerge. Leaves pinnate; leaflets 3–7 pairs, sometimes staggered on the rhachis to give the appearance of a terminal leaflet, the terminal and upper leaflets equal to or longer than the lower, entire or crenate, with simple hairs or glabrous. Inflorescences terminal or in the axils of young or undeveloped leaves, with sessile or stalked dense clusters of shortly pedicelled flowers. Sepals (3–)4–5, joined in the lower part, herbaceous, hairy outside; petals absent. Disk complete, flat or cup-shaped, glabrous. Stamens 4–5; filaments glabrous, in bud bent through two zigzags; anthers basifixed, spherical in outline, the connective elliptic, as wide as the anther-sacs; staminodes lacking anthers. Ovary 2-locular; style with 3 stigmatic lobes; ovules 2 per locule, suspended from the upper part of the axis. Fruit a reddish drupe with one cavity and one seed by abortion. Aril absent; testa of seed with stomata.

A genus of 3 species in tropical Africa plus one in Madagascar.

Leaf-rhachis and underside of leaflets glabrescent; leaflet apex mostly acuminate; fruit 1.4–2 cm. long, glabrous; clusters of male flowers on 3–10 cm. peduncles . . . . 1. *Z. golungensis*

Leaf-rhachis and underside of leaflets ferruginous pubescent; leaflet apex obtuse; fruit up to 2.6 cm. long, velvety; clusters of male flowers ± sessile on the young twigs, usually precocious (but can have stalks 0.5–3 cm. long), often forming dense masses . . . . 2. *Z. africana*

1. **Z. golungensis** *Hiern*, Cat. Afr. Pl. Welw. 1: 128 (1896); Radlk. in E.P. IV, 165: 1421 (1933); T.T.C.L.: 561 (1949); I.T.U., ed. 2: 384 (1952); F.P.S. 2: 343 (1952); C.F.A. 2, 1: 92 (1954); F.W.T.A., ed. 2, 1: 725 (1958); Hauman in F.C.B. 9: 361 (1960); F.F.N.R.: 225 (1962); Haerdi in Acta Trop., suppl. 8: 125 (1964); Exell in F.Z. 2: 539, t. 115B (1966); Fouilloy & Hallé in Fl. Cameroun 16: 189 (1973) & in Fl. Gabon 23: 189 (1973); Hamilton, Field Guide Uganda For. Trees: 223 (1981); Troupin, Fl. Rwanda 2: 310, fig. 97/2 (1983); Vollesen in Fl. Ethiopia 3: 493, fig. 125.3 (1990); Beentje, K.T.S.L.: 424, fig. (1994). Type: Angola, Golungo Alto, Quibola, *Welwitsch* 4545 & 4546 (LISU, syn., K!, isosyn.)

Tree 6–24 m. tall; bark grey-brown, smooth, flaking in irregular circular patches, covered in reddish warty lenticels and petiole-scars. Leaf-axis 15–25 cm. long, terete (winged when young), sparsely hairy, glabrescent; petiolules 0–2 mm. long; leaflets in 4–6 pairs, mostly elliptic, varying slightly towards ovate and elliptic-oblong, shortly and obtusely acuminate, 6–11(–17) cm. long, 2–4(–5.5) cm. wide (the lower scarcely shorter than the upper), entire or crenate, glabrous; lateral nerves in 12–16 pairs, rather regularly and closely spaced, looped within the margins. Male flowers olive green or yellowish, in dense cymose clusters 1–2.5 mm. long on stalks 2–10 mm. long at the ends of the peduncles, pedicels 0–2 mm. long; female flowers in a sessile cluster or short panicle, pedicels 1–3 mm. long, densely pubescent, lengthening in fruit.

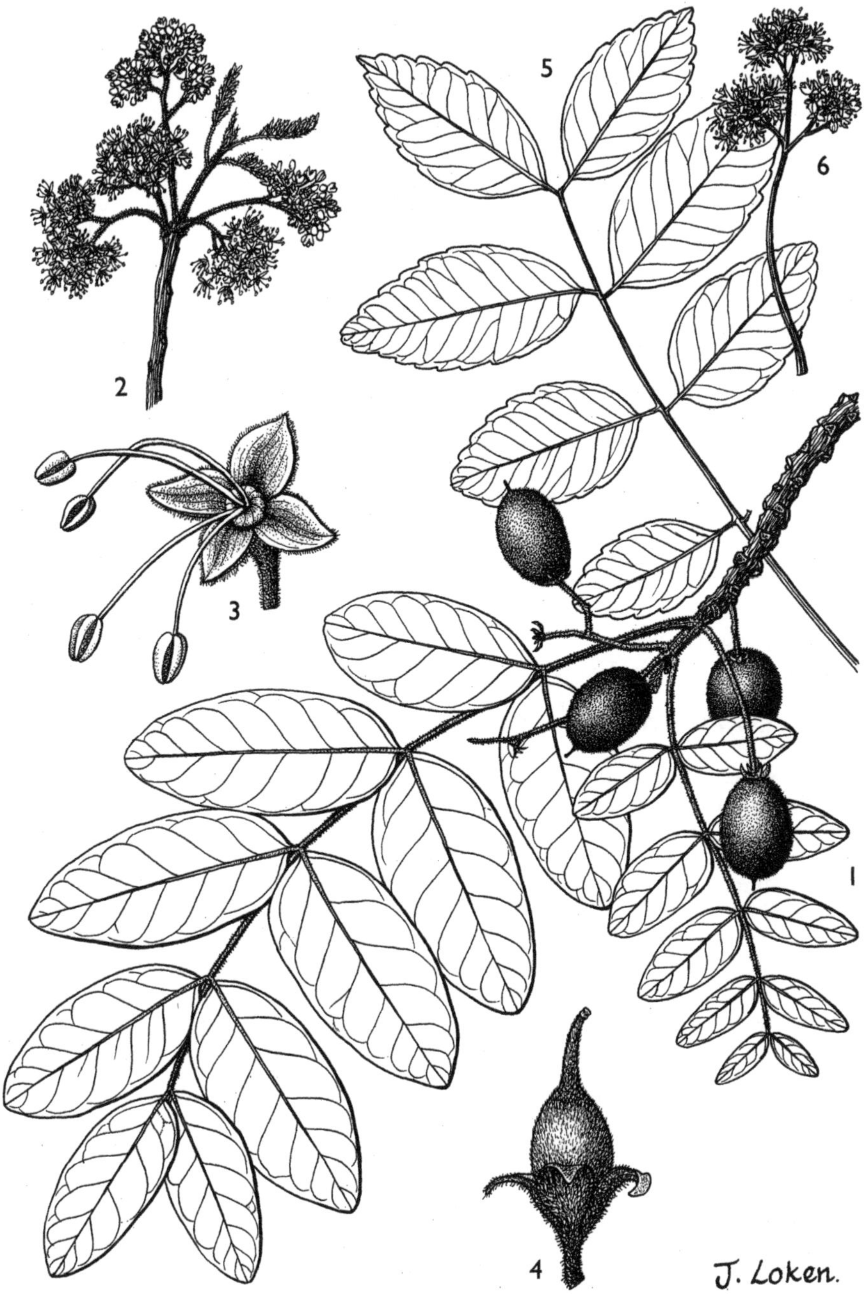

FIG. 3. *ZANHA AFRICANA* — **1**, branchlet with fruit, × $^2/_3$; **2**, inflorescences and young leaves, × $^2/_3$; **3**, male flower, × 4; **4**, young fruit, × 4. *ZANHA GOLUNGENSIS* — **5**, leaf, × $^2/_3$; **6**, inflorescence, × $^2/_3$. 1, from *Chase* 950 & *Angus* 1786; 2, 3, from *Chase* 951; 4, from *Chase* 947; 5, from *Chase* 1540, 6, from *Chase* 536. Reproduced from F.Z. 2, t. 115 (1966).

Sepals olive green, erect, ± 4.3 mm. long, united to ± half-way; disk flat, green and glandular when fresh, dark red in dried specimens. Stamens 4; filaments 2–7 mm. long; after anthesis the filaments drop off the receptacle leaving a star-shaped configuration of sepals. Ovary rudiment absent. The male inflorescence drops off as a single unit after flowering. Infructescence a lax thyrse; drupe pink, yellow or orange, olive-like, 1.4–2 cm. long, glabrous, with persistent sepals and style. Fig. 3/5–6.

UGANDA. Acholi District: Paimol, Mar. 1935, *Eggeling* 1769!; Bunyoro District: Budongo Forest, Feb. 1935, *Eggeling* 1631!; Teso District: Serere, Feb. 1933, *Chandler* 1064!

KENYA. Meru District: Lower Imenti Forest, 26–28 June 1974, *R.B. & A.J. Faden* 74/901B; Kwale District: Shimba Hills National Reserve, Makadara Forest, 13 Jan. 1972, *R.B. & A.J. Faden* 72/71!

TANZANIA. Mbulu District: Lake Manyara National Park, Mto wa Ukindu, 22 Nov. 1963, *Greenway & Kirrika* 11068!; Tanga District: Mlinga peak, 2 Feb. 1939, *Greenway* 5855!; Morogoro District: Turiani, Nov. 1954, *Semsei* 1956!

DISTR. **U** 1–4; **K** 4 (*fide* EA), 7; **T** 2–4, 6, 8; Senegal to Ethiopia and south to Angola, Zambia, Malawi, Mozambique and Zimbabwe

HAB. Deciduous woodland, commonly associated with *Combretum* and *Brachystegia*, often on rocky hills, also in rain-forest; 300–1700 m.

NOTE. The timber is said to be useful for furniture and building; it carves easily into small articles. Eggeling says that the fruit is edible and has an orange aril; this is apparently the fleshy mesocarp. It is eaten by chimpanzees. Other reports say that the seed is poisonous. The bark has been noted on herbarium sheets as a remedy for chest complaints and also as a soap substitute.

2. **Z. africana** (*Radlk.*) *Exell* in F.Z. 2: 537, t. 115A (1966); Verdc. & Trump, Common Pois. Pl. E. Afr.: 108 (1969); Vollesen in Opera Bot. 59: 58 (1980); Beentje, K.T.S.L.: 423 (1994). Lectotype, chosen here: Tanzania, Songea, Ungoni, Kwa Mtumbo, *Busse* 785 (EA!, lecto., EA!, K!, isolecto.)

Small tree 3.5–12 m. tall with rather nodular gnarled grey branches; bark red, brittle, reticulate, flaking off in patches, with orange layer beneath. Leaves up to ± 23 cm. long; petiole 1–5 cm. long, ferruginous pubescent; leaf rhachis 4–35 cm. long, similarly pubescent; leaflets in 3–8 pairs, often not exactly opposite, ovate, elliptic or almost round, the lower pairs considerably smaller than the upper; upper leaflets 8–15 cm. long, 4–8 cm. wide, obtuse at the apex, truncate to cordate at the base, crenate towards the apex or rarely entire, glabrous to sparsely or densely ferruginous pubescent; lateral nerves in up to 14 pairs, the venation prominently reticulate beneath; petiolules 1–3 mm. long. Male inflorescences precocious, ± sessile clusters, 1.5 cm. wide, forming masses ± 5 cm. wide, peduncles sometimes 0.5–?3 cm. long; female inflorescences tawny felted, with peduncle up to 2 cm. long, lengthening in fruit. Flowers greenish, sweet-scented; sepals 5, 4 mm. long, united for 1 mm. at base; disk cup-shaped, 0.7 mm. high. Stamens and staminodes 5; filaments of stamens 9 mm. long; anthers 1.2 mm. long. Fruit bright orange, ellipsoid, up to 2.6 cm. long, 1.3–1.8 cm. wide, velvety; sepals and style base persistent; stalks 0.3–1.9 cm. long. Fig. 3/1–4.

KENYA. Machakos District: Makueni, 24 Nov. 1955, *Glover* in *E.A.H.* 14/56!

TANZANIA. Mwanza District: Geita, Sept. 1949, *F.H.* 2852!; Ufipa District: ridge E. of Lake Kwela, 7 Nov. 1950, *Bullock* 3473!; Kilosa District: between Berega and Mlali on Mpwapwa road, 10 Dec. 1935, *B.D. Burtt* 5384!

DISTR. **K** 4; **T** 1, 2, 4–8; Zaire, S. Angola, Zambia, Malawi, Mozambique and Zimbabwe

HAB. Deciduous woodland, commonly associated with *Combretum* and *Brachystegia*, often on rocky hills; 300–1750 m.

SYN. *Dialiopsis africana* Radlk. in E. & P. Pf., Nachtr. 3: 207 (1907) & in E.P. IV, 165: 1419 (1933); T.T.C.L.: 557 (1949); K.T.S.: 509 (1961); F.F.N.R.: 224 (1962); Haerdi in Acta Trop., suppl. 8: 124 (1964)

NOTE. The fruits contain ± 10.5% saponins and are used as soap after boiling; reports from South Africa suggest they are edible. The bark of the roots appears to contain powerful drug elements. It is used in childbirth and against fungal infections, fits, insanity, pains of head and neck and other complaints, both by rubbing into incisions or on to the skin and taken internally. Fatalities following ingestion of an infusion of pounded root are known.

The syntypes of *Z. africana*, *Busse* 785 and 785a, Rovuma R., Kwa-Ktira (seeds only) were destroyed at Berlin; one of the isosyntypes at EA, *Busse* 785, has therefore been selected as lectotype.

4. MAJIDEA

Oliv.* in Hook., Ic. Pl. 11, t. 1097 (1871); Radlk. in E.P. IV, 165: 1462 (1934)

Harpullia Roxb. subgen. *Majidea* (Oliv.) Radlk. in Act. Congr. Bot. Amst. 1877: 123 (1879)

Monoecious tall trees with greyish bark; twigs, leaves and inflorescences initially densely fasciculate-hairy. Leaves with 4–11 pairs of leaflets; rhachis 4-angled, with lines of simple hairs on the angles; leaflets ovate to lanceolate, slightly asymmetric, entire, glabrescent. Inflorescence terminal, paniculate, flowers subtended by herbaceous bracts. Flowers regular; sepals 5, free, concave, herbaceous, externally hairy; petals 4, caducous, without claw or scale; disk a complete circle. Stamens 7–8(–10); filaments folded in bud, glabrous; anthers sagittate. Ovary 3-lobed, 3-locular with 2 ovules in each locule; style long, hairy, with a glabrous channelled tip. Fruit a dehiscent trigonous capsule, opening to reveal 3 bright pink valves. Seeds 1–2 per valve, ovoid, without aril, softly velvety, blue-black; cotyledons spirally arranged.

A genus confined to tropical Africa with 2–3 species in mainland Africa and Madagascar.

Young branches pale to mid-brown with conspicuous pale lenticels; capsule brownish outside, pink inside; leaflets 3–6 cm. long (coastal) . . . . 1. *M. zanguebarica*
Young branches dark brown, lenticels not noticeable; capsule reddish outside as well as inside; leaflets 7–9 cm. long (**U** 2, 4; **T** 1) . . . . 2. *M. fosteri*

1. **M. zanguebarica** *Oliv.** in Hook., Ic. Pl. 11, t. 1097 (1871); Radlk. in E.P. IV, 165: 1466 (1934); T.T.C.L.: 559 (1949); K.T.S.: 515 (1961); Capuron in Mém. Mus. Nat. Hist. Nat., Paris, 19: 43, t. 8 (1969); Vollesen in Opera Bot. 59: 58 (1980); Noad & Birnie, Trees of Kenya, 263, fig. on 264 (1989); Beentje, K.T.S.L.: 421, fig. (1994). Type: Tanzania, Uzaramo District, Dar es Salaam, *Kirk* 118 (K!, syn. – 3 sheets)

Shrub or tree 2–22 m. tall; bark rough or ± smooth, whitish buff; slash cream to pale orange; twigs with raised whitish lenticels. Leaf-rhachis 10–30 cm. long, 4-angled in cross-section; petiolules 1 mm. long; leaflets drying bright green, drooping when young, in 5–7 pairs, arranged regularly or tending to be alternate, ovate-lanceolate, 5–7 cm. long, 2–3 cm. wide, shortly bluntly acuminate or acute, asymmetrically cuneate at the base, glabrous and shining; lateral veins in 7–15 faint pairs. Inflorescence with main axes 8–15(–30) cm. long, all parts densely velvety pubescent; flowers pleasantly scented; bracts elliptic to lanceolate-elliptic, 1.3–1.7 cm. long, 6–8 mm. wide, pedicels 3–6 mm. long. Sepals ovate, 5–7 mm. long, the outermost larger

* Usually attributed to Kirk and certainly the name must have been suggested by Kirk after the Sultan of Zanzibar. I have found no Kirk letters in the archives concerning this. It can of course be given as Kirk ex Oliv. but that format is not used in this Flora. [B.V.]

than the other 4; petals 4, yellow-green to red with white hairs, elliptic, 6 mm. long, 2 mm. wide, soon falling; disk hardly visible in the female flowers, pink-tinged and conspicuous in the male. Stamens exserted; filaments yellow-green to orange-red, 6 mm. long; staminodes very short. Male flowers without an ovary rudiment. Fruit a thinly woody trigonous apiculate capsule, externally brownish yellow, internally bright pink, 3–3.5 cm. long, 3.5–4.5 cm. wide. Seeds bluish black, ovoid, 1.2 cm. long, finely silky pubescent, ± persistent. Mature fruits and male flowers commonly occur together in the inflorescence. Fig. 4.

subsp. **zanguebarica**; Capuron in Mém. Mus. Nat. Hist. Nat., Paris, 19: 43, t. 8/14–20 (1969)

Capsules densely finely tomentose with persistent stellate hairs.

KENYA. Kwale District: Diani Forest, 11–13 July 1972, *Gillett & Kibuwa* 19874!; Kilifi District: Marafa, 19 Nov. 1961, *Polhill & Paulo* 803!; Lamu District: Witu, Mambosasa Forest Station, 29 Jan. 1958, *Verdcourt* 2128!

TANZANIA. Tanga District: near Moa, Mtotohovu, 10 Sept. 1951, *Greenway* 8708!; Pangani District: Bushiri, 30 Nov. 1950, *Faulkner* 715!; Uzaramo District: Dar es Salaam, Yacht Club, 4 Jan. 1970, *B.J. Harris* 3806!; Zanzibar, Pangajuu Cave-well, 10 June 1930, *Vaughan* 1340!

DISTR. **K** 7, **T** 3, 6, 8; **Z**; Madagascar; cultivated in Nairobi Arboretum, also in Rodrigues, India (Assam), Sri Lanka, Singapore and Hawaii

HAB. Dry evergreen forest fringes, riverine and coastal bushland; particularly *Julbernardia, Cynometra, Brachystegia* woodland, lowland evergreen forest, semideciduous often flooded forest, open ground with *Hyphaene* etc., also abandoned sisal plantations, often on coral; 0–300 m.

SYN. *Harpullia zanguebarica* (Oliv.) Radlk. in Sitz. Akad. Wiss. München 20: 279 (1890)

NOTE. Capuron (1969) has included *M. madagascariensis* (Baill.) Radlk. in E.J. 56: 256 (1920) (*Cossignia madagascariensis* Baill. in Adansonia 11: 247 (1874)) as a subspecies of *M. zanguebarica* and this appears to be acceptable. The timber is strong and white but not durable. [I have not seen any **T** 8 material but Davies presumably did so – B.V.]

2. **M. fosteri** (*Sprague*) *Radlk.* in E.J. 56: 255 (1920) & in E.P. IV, 165: 1463 (1934); I.T.U., ed. 2: 381 (1952); F.P.S. 2: 341 (1952); Wimbush, Cat. Kenya Timbers, ed. 2: 53 (1957); F.W.T.A., ed. 2, 1: 725 (1958); Hauman in F.C.B.: 365 (1960); Fouilloy & Hallé in Fl. Cameroun 16: 195 (1973) & in Fl. Gabon 23: 195, t. 44/5–9 (1973); Hamilton, Field Guide Uganda For. Trees: 222, fig. 409 (1981). Type: Nigeria, Lagos, Oloke Meji, *Foster* 49 (?K, not found)*

Tree 15–35 m. tall with a broad sometimes buttressed trunk; bole clean up to 18 m., 2 m. girth; bark grey, grey-brown or yellowish, thin, smooth, densely lenticulate, slightly flaking, said to smell of thymol; slash creamy yellow or yellow-brown. Leaf-rhachis 15–20 cm. long; leaflets in 5–9 pairs, dull green tending to olive when dry, lanceolate or oblong-lanceolate, 5–10.5 cm. long, 1.5–3 cm. wide, tapering to a narrow, usually acuminate tip, glabrous. Inflorescence-axes overall mid-olive to brown, up to 13 cm. long, densely covered in fasciculate hairs. Flowers green, tinged pink. Disk of male flowers very conspicuous, convex, broad, red, drying black. Fruit reddish outside, bright pink or scarlet inside, 3.5–4 cm. long, 5 cm. wide. Seeds purplish, 1.4 cm. across, velvety.

UGANDA. Bunyoro District: Budongo, Nov. 1932, *C.M. Harris* 155! & Bugoma Forest Reserve, 17 Oct. 1962, *Styles* 155!; Mengo District: Mabira Forest, 1.5 km. NE. of Najembe, 30 June 1950, *Dawkins* 602!

TANZANIA. **Bukoba, Sept. 1950, *Watkins* 537

* Gillett noted it was missing in 1950; Radlkofer saw it at Kew pre-1927.

** Must also occur in S. Minziro Forest, Bukoba District, since it is in Ugandan part of forest.

FIG. 4. *MAJIDEA ZANGUEBARICA* — **1**, flowering branchlet (inflorescence somewhat simplified), × $^2/_3$; **2**, male flower (one petal removed), × 4; **3**, female flower (one sepal and one petal removed), × 4; **4**, fruit, × $^2/_3$; **5**, valve of fruit with seed, × $^2/_3$. 1, from *Verdcourt* 2128; 2, from *Faulkner* 715; 3, from *Greenway & Rawlins* 9371; 4, 5, from *Greenway* 8708. Drawn by Pat Halliday.

DISTR. **U** 2, 4; **K** 5 (*fide* Wimbush); **T** 1 (*fide* EA); Ivory Coast to Central African Republic and Sudan

HAB. Rain-forest, closed high forest; 1000–1200 m.

SYN. *Harpullia fosteri* Sprague in K.B. 1908: 434 (1908); I.T.U.: 216 (1940)
H. multijuga Radlk. in Z.A.E.: 481 (1912) & in E.P. IV, 165: 1464 (1934). Type: Congo, Ituri, between Awakubi and Bomili, near Bawaka, *Mildbraed* 3245 (B†, holo.)
Majidea multijuga (Radlk.) Radlk. in E.J. 56: 255 (1920)

5. **APORRHIZA**

Radlk. in Sitz. Akad. Wiss. München 8: 338 (1878) & in E.P. IV, 165: 1134 (1933)

Small to medium-sized, monoecious trees. Leaves with 2–7(–11) pairs of entire coriaceous leaflets. Inflorescence terminal or axillary, paniculate. Flowers regular; sepals 5, nearly free, densely pubescent; petals 5, clawed, with a small to equally large bilobed scale, shaggy-hairy in the lower half; disk glabrous, lobed. Stamens (6–)7–8; filaments folded in bud; anthers oblong, sagittate at the base, glabrous. Ovary bilobed, bilocular; style short. Fruit with 2 (or 1 by abortion) diverging compressed dehiscent mericarps, externally pubescent with either grey, green or golden hairs, glabrous and reddish inside. Seeds black, bean-shaped, more than half covered by a yellow or orange aril.

About 6 species in tropical Africa, poorly known apart from *A. paniculata*; there is a shortage of flowering material in herbaria.

A. paniculata *Radlk.* in Sitz. Akad. Wiss. München 8: 339 (1878) & in E.P. IV, 165: 1135, fig. 31 (1933); F.P.S. 2: 336 (1952); Hauman in F.C.B. 9: 319 (1960); Verdc. in Journ. E. Afr. Nat. Hist. Soc. 24 (106): 65 (1963); Fouilloy & Hallé in Fl. Cameroun 16: 157 (1973) & in Fl. Gabon 23: 157 (1973) (but not occurring in Cameroon or Gabon); Beentje, K.T.S.L.: 414, fig. (1994). Type: Sudan, Nabambissoo, *Schweinfurth* 3041 (B†, holo., K!, P, iso.)

Tree 4.5–22 m. tall; bark smooth, grey; young parts and fruits with a short, velvety, reddish or grey-green pubescence. Petiole 1–9 cm. long, woody; rhachis 10–36 cm. long; leaflets in 3–5 pairs, subopposite; petiolules fleshy, sparsely pilose; blade oblong-elliptic or elliptic, the terminal pair largest, 9–25 cm. long, 5–9 cm. wide, the other pairs progressively smaller, shortly and bluntly acuminate, the base acute, coriaceous, glabrous throughout or some hairs on the veins beneath; lateral veins in 8–11 pairs, prominent beneath, impressed above. Inflorescence-axes 20–50 cm. long; branches 2–16 cm. long. Calyx-lobes green, 2 mm. long, silky hairy; petals yellow or white, equalling the sepals, claw 1 mm. long, limb 1.5 mm. long, ovate, scale bilobed, ciliate. Stamens 7–8; filaments bent into a zigzag in bud, pilose throughout, 2.8 mm. long at maturity. Ovary rudiment in male flowers similar to but much smaller than the ovary. Mericarps 2, semicircular in outline, compressed, 1.5–2 cm. long, 1–1.2 cm. wide, densely finely adpressed-pubescent, internally glabrous and shining. Seed black, ± 1.3 cm. long, 5 mm. wide, $^4/_5$ covered by an orange aril. Fig. 5.

KENYA. Kwale District: Shimba Hills, Sheldrick's Falls, 2 Apr. 1968, *Magogo & Glover* 615! & Marere pumping station, 3 Apr. 1968, *Magogo & Glover* 753!; Tana R., plot 15B, 24 Nov. 1982, *F. Hughes*!

TANZANIA. Handeni District: R. Chazi on Turiani–Mahonda road, 4 Nov. 1947, *Parry* in *F.H.* 2208!; Mpanda District: 16 km. S. of Mpanda, Sept. 1961, *Procter* 1941!; Uzaramo District: Vikindu Forest Reserve Aug. 1953 *Paulo* 123!; Zanzibar, Jozani Forest Reserve, 29 Mar. 1981, *Mturi* 367!

DISTR. **K** 7; **T** 3, 4, 6, 7; **Z**; Nigeria, Central African Republic, Zaire, Sudan, Zambia, Malawi, Mozambique and Zimbabwe

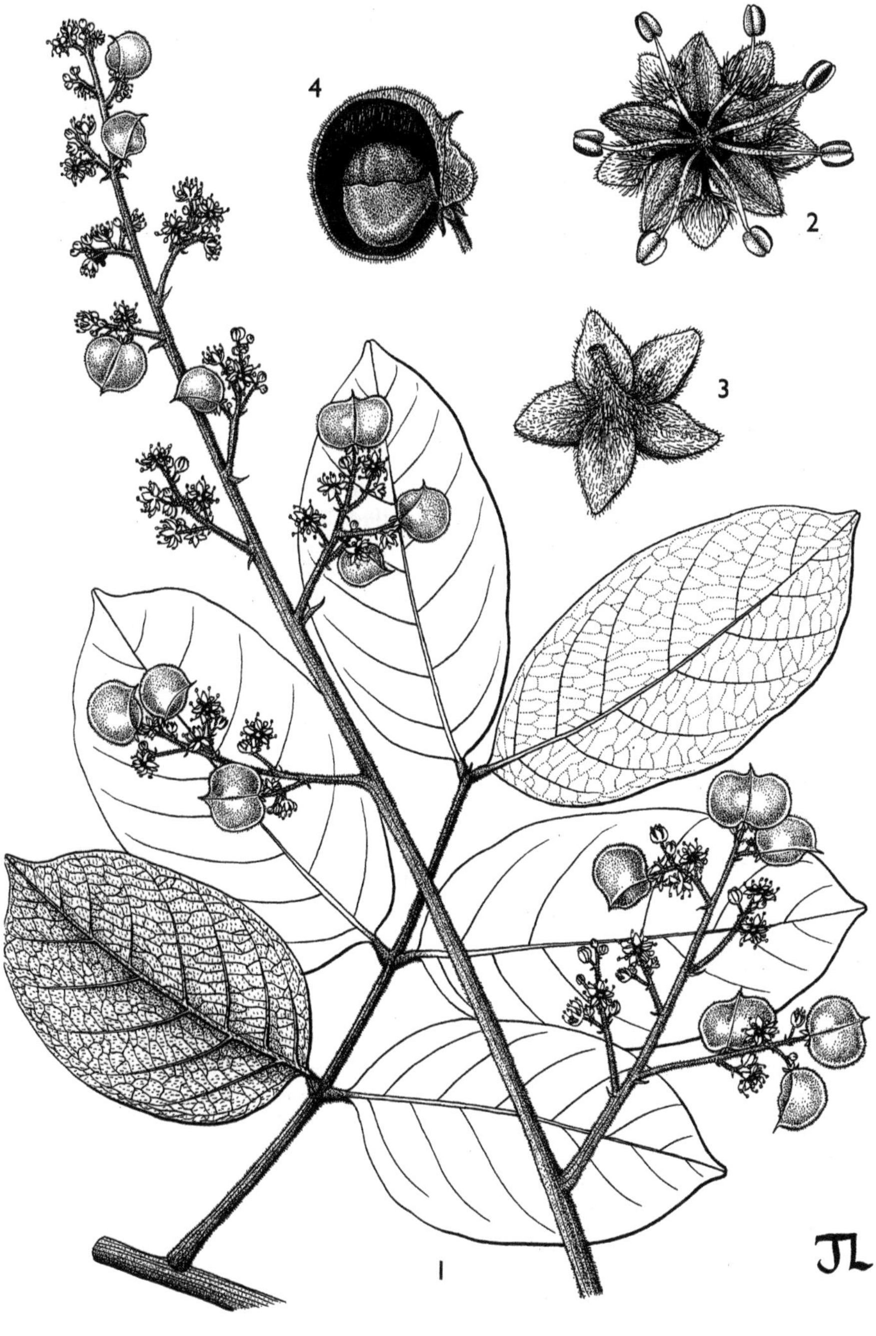

FIG. 5. *APORRHIZA PANICULATA* — **1**, leaf and inflorescence, × $^{2}/_{3}$; **2**, male flower, × 6; **3**, calyx, × 6; **4**, one coccus of fruit opened to show seed, × 2. 1–3, from *Grout* 10/49; 4, from *Duff* 242/37. Reproduced from F.Z. 2, t. 103 (1966).

HAB. Riverine forest, sometimes growing in the water, ground-water forest, sometimes 'between rocks' (*fide* Haerdi); 20–1200 m.

SYN. *A. nitida* Gilg in V.E. 3(2): 280 (1921); Milne-Redh. in K.B. 1931: 272 (1931); Radlk. in E.P. IV, 165: 1136 (1933); T.T.C.L.: 556 (1949); F.W.T.A., ed. 2, 1: 721 (1958); Hauman in F.C.B. 9: 318 (1960); F.F.N.R.: 224 (1962); Haerdi in Acta Trop., suppl. 8: 123 (1964); Exell in F.Z. 2: 514, t. 103 (1966); Fouilloy & Hallé in Fl. Cameroun 16: 156 (1973) & in Fl. Gabon 23: 156 (1973). Type: Malawi, Blantyre, *Buchanan* in *Medley Wood* 6992 (B†, holo., K!, iso.)

NOTE. *A. nitida* and *A. paniculata* were separated on the colour of the indumentum, being rusty or grey-green but it varies and is not correlated with any other character. Verdcourt appears to have been the first to combine them. The authority of *A. nitida* is usually given as Gilg ex Milne-Redh. but there appears to be enough in Gilg's description to validate the name.

6. LYCHNODISCUS

Radlk. in Sitz. Akad. Wiss. München 8: 271, 332 (1878) & in E.P. IV, 165: 1137 (1933)

Small or medium trees; young parts rusty-tomentose. Leaves pinnate; leaflets entire or more commonly toothed, at least in the upper part, the middle leaflets a little larger than the others. Inflorescence an axillary or terminal paniculate thyrse; flowers regular, unisexual, in clusters subtended by bracts, each inflorescence successively male, female and finally male. Sepals 5, valvate, joined at the base, densely hairy outside; petals 5, clawed, widest just below the apex, with a scale which is reflexed and bearded at the tip; disk circular, glabrous, the outer part flat, lobed, the inner a raised rim, together resembling a cup and saucer. Stamens 10–13; filaments folded in bud, glabrous or hairy; anthers oblong, dorsifixed. Ovary 3-locular; style undivided. Fruit a 3-valved dehiscent ellipsoid or obovoid capsule, externally hairy, internally glabrous. Seeds 3, trigonous in cross-section, enclosed almost totally by a waxy aril.

Six species in tropical Africa, mainly in the west, one species extending to Uganda.

L. cerospermus *Radlk.* in E. & P. Pf. III, 5: 344 (1895); & in E.P. IV, 115: 1140 (1933); I.T.U., ed. 2: 380 (1952); Hauman in F.C.B. 9: 326, fig. 6 (1960); Fouilloy & Hallé in Fl. Cameroun 16: 168 (1973) & in Fl. Gabon 23: 168 (1973); Hamilton, Field Guide Uganda For. Trees: 223, fig. 412 (1981). Type: Zaire, Kasai, Mukende, *Pogge* 703 (B†, holo., ?COI, iso.)*

Slender tree 3–20 m. tall (or, according to Eggeling, partly scandent); bark smooth, greyish white with patches of orange; young parts chocolate- or rusty-tomentose with spreading to ascending hairs. Petiole up to 10 cm. long; rhachis 10–20 cm. long; leaflets in 6–7 pairs, opposite or alternate; petiolules 3–6 mm. long, fleshy, hairy; blade oblong-elliptic, ovate-oblong or the lowermost pair of leaflets elliptic, middle leaflets largest, 11–15 cm. long, 3–5 cm. wide, the uppermost a little smaller, the lowermost 6–10 cm. long, 3–5 cm. wide, apex with a sharp acumen 5–25 mm. long or acute, toothed throughout or only towards the apex, the teeth 1–2 mm. deep, or the margin merely sinuate, glabrous apart from the midrib and some or all veins beneath; lateral veins in 10–16(–22) pairs, curving just inside the margin and terminating in the teeth; midrib impressed above, prominent beneath. Inflorescences golden brown, axillary and terminal, the main axes 14–45 cm. long, the main branches 5–10 cm. long; flowers grey-white in groups of 1–4 subtended by

* Meikle has annotated *Eggeling* 535 and *Fyffe* 62 as '*L. cerospermus* Radlk. compared with type Oct. 1948' but does not specify from which herbarium it came. The E.P. description shows that Radlkofer saw only material at Berlin.

FIG. 6. *LYCHNODISCUS CEROSPERMUS* — **1**, **2**, flowering branchlet (with young inflorescence), × $^{2}/_{3}$; **3**, male flower, × 6; **4**, male flower (one sepal and two petals removed), × 6; **5**, **6**, petal (male flower), × 6; **7**, fruit, × 1; **8**, fruit (apical view), × 1; **9**, **10**, seed, × 1. 1, 2, from *Eggeling* 3738; 3–6, from *Eggeling* 1140; 7–10, from *Chandler* 2423. Drawn by Pat Halliday.

4–6 mm. long lanceolate bracts; pedicels 4–6 mm. long. Sepals 2.5 mm. long, spreading flat in the male flower, glabrescent on the inner surface; petals funnel-shaped, 2.8–3.5 mm. long, 3–3.5 mm. wide, white, the scale equalling the limb; disk 1 mm. high. Stamens 13, glabrous throughout; filaments 4 mm. long. Ovary 3-lobed; style entire with 3 stigmatic grooves; ovary rudiment in male flowers a small cushion of hairs. Fruiting pedicels usually enlarging, 0.4–1.3 cm. long; capsule bright red when ripe, 3-lobed with rounded angles, 1.6–2 cm. long and wide, externally with some coarser hairs amongst the fine tomentum. Seeds dark, (1-)3, 1.3–1.8 cm. long, 1–1.3 cm. wide; aril red. Fig. 6.

var. **cerospermus**

Leaflets toothed throughout the margin.

UGANDA. Bunyoro District: Budongo Forest, Mar. 1933, *Eggeling* 1140!; Toro District: Fort Portal, Kibale Forest, Aug. 1936, *Eggeling* 3142!; Masaka District: Minziro Forest, July 1938, *Eggeling* 3738!

DISTR. **U** 2, 4; Zaire

HAB. Understorey of forest; 1100–1350 m.

var. **mortehanii** (*De Wild.*) *Hauman* in F.C.B. 9: 328 (1960); Fouilloy & Hallé in Fl. Cameroun 16: 169 (1973) & in Fl. Gabon 23: 169 (1973). Lectotype, chosen by Hauman: Zaire, Dundusana, *Mortehan* 42 (BR, lecto.)*

Leaflets entire or toothed only at apex.

UGANDA. Bunyoro District: Budongo Forest, Feb. 1938, *Eggeling* 3530!; Toro District: N. Ruwenzori Forest Reserve, 26 July 1960, *Paulo* 605!; Masaka District: Malabigambo Forest, *Fyffe* 62!

DISTR. **U** 2, 4; Zaire and Sudan

HAB. Wet and dry evergreen forest; 1050–1500

SYN. *L. mortehanii* De Wild. in F.R. 13: 378 (1914)
Laccodiscus sp. nov. sensu I.T.U.: 217 (1940)

NOTE. The distinction between the two varieties is such a trivial one that it may prove impossible to maintain, despite being kept up by authors of Floras of nearby countries. There is no geographical correlation.

Dale & Eggeling (I.T.U., ed. 2: 380 (1952)) also record this species from **U** 1, W. Nile District, Madi, Zoka Forest, but it is not known to which variety it belongs.

7. **BLIGHIA**

Koenig in Ann. Bot. 2: 571 & t. 16, 17 (1806); Radlk. in E.P. IV, 165: 1142 (1933); Wilczek in B.J.B.B. 21: 149–165 (1951)

Phialodiscus Radlk. in Sitz. Akad. Wiss. München 9: 479 (1879) & in E.P. IV, 165: 1147 (1933)

Trees or large shrubs. Leaves pinnate; leaflets in 1–5 pairs, tending to be opposite, entire. Inflorescences axillary, thyrsoid, tending to be reduced to racemes. Flowers regular, subtended singly or in small groups by minute bracts; sepals 5, almost free or joined up to halfway, opening in bud to reveal the immature anthers; petals 5, scale equal to or smaller than the limb and joined laterally to it to form a funnel. Disk circular, glabrous or hairy. Stamens (6–)8–10;

* Hauman cites *Mortehan* 42 as holotype but De Wildeman cited another specimen, *de Georgi* 1076 also; I have taken Hauman's choice as a lectotypification.

filaments straight and very short in bud, elongating as the flower matures; anthers oblong, pilose, glabrescent, the base sagittate; staminodes similar but the filaments remain short. Ovary 3(–4)-lobed, golden-hairy; style short, blunt, with 3(–4) furrows. Ovary rudiment in male flowers a minute white-pilose cushion. Fruit a glabrous, 3(–4)-gonous, turbinate capsule, splitting into 3(–4) valves, with glabrous or velvety-pubescent inner surfaces. Seeds ovoid, glossy, with a yellow cup-shaped aril around the basal part.

Three species in tropical Africa, 2 of which are native in East Africa, and one which is cultivated.

Blighia sapida Koenig (see F.W.T.A., ed. 2, 1: 722, fig. 198 (1958)) the akee, a native of West Africa, has been grown in Nairobi, near the Scott Agricultural Laboratories, 8 June 1980, *Gillett* 22829. Monoecious tree 7.5–15 m. tall with pinnate leaves to ± 25 cm. long; leaflets in (2–)3–6 pairs, elliptic, oblong, ovate or obovate, 3–18 cm. long, 2–8.5 cm. wide. Flowers greenish white, very fragrant, small, in axillary racemes; petals 5; stamens 8–10. Fruit yellow then bright red, oblong-ovoid, ± 3 angled, 4–7.5 cm. long, 3–5 cm. wide, ± 3 lobed at the apex, splitting when ripe to reveal 3 large shiny black seeds in a creamy white or yellow aril. Although widely grown throughout the tropics as an edible fruit care must be taken since it is poisonous both when unripe and when overripe; the seeds are poisonous. The unripe fruits contain a poisonous non-protein amino-acid hypoglycin A which causes a lowering of the blood sugar level sometimes resulting in death.

Fruit small, not exceeding 3 cm., valves glabrous inside; leaflet midrib with a raised central line along at least part of its length on the upper leaf surface · · · · · · · · · · · · · · · · · · · 1. *B. unijugata*
Fruit large, exceeding 3 cm., valves furry inside; leaflet midrib in a channel on the upper leaf surfaces:
Fruit with sharp or winged angles; petiole narrowly winged; leaflets usually bluntly acuminate, paler beneath · · · · · · 2. *B. welwitschii*
Fruit with blunt, square angles; petiole flattened but not winged; leaflets usually obtuse but sometimes acute or shortly acuminate, the same colour above and beneath (cultivated) *B. sapida*

1. **B. unijugata** *Baker* in F.T.A. 1: 427 (1868); R. Wilczek in B.J.B.B. 21: 159, fig. 41 (1951) F.P.S. 2: 336 (1952); F.W.T.A., ed. 2, 1: 723 (1958); Hauman in F.C.B. 9: 316 (1960); K.T.S.: 507 (1961); F.F.N.R.: 224 (1962); Haerdi in Acta Trop., suppl. 8: 123 (1964); Exell in F.Z. 2: 516, t. 104 (1966); Fouilloy & Hallé in Fl. Cameroun 16: 186 (1973) & in Fl. Gabon 23: 186, t. 43/8–15 (1973); Hamilton, Field Guide Uganda For. Trees: 225, fig. 419 (1981); Troupin, Fl. Rwanda 2: 302 (1983); Vollesen in Fl. Ethiopia 3: 508, fig. 125.14 (1990); Beentje, K.T.S.L.: 414, fig. (1994). Type: Cameroon, Ambas Bay, *Mann* 760 (K!, holo.)

Tree, probably dioecious, usually 6–9 m. but sometimes reaching 30 m. tall, with a dense crown; bark grey-brown smooth, thin, with warts and horizontal ridges; slash white, pale red or brown with white streaks; young twigs golden-hairy, glabrescent, dark grey. Leaves reddish when young, drying bright green or brownish; petiole up to 4 cm. long; rhachis 0.6–10 cm. long; petiolules 1–2 mm. long; leaflets in (1–)2–4(–5) pairs, elliptic, oblong, obovate or oblanceolate, the upper pair 3–26 cm. long, 1.2–9.5 cm. wide (but in West Africa up to 30 × 13 cm.), the lowest either similar (usually when there are only 2 pairs) or as small as 2 × 1.2 cm., acuminate at apex and tapering or sometimes rounded at the extreme tip, the midrib visible as a raised line on the upper surface, glabrous apart from some hairs on the veins plus tufts in the nerve axils; lateral veins in 6–10(–12) pairs, curving and ending well inside the leaflet margin, usually whitish beneath, glabrous. Inflorescences borne in axils of current leaves, 5–10 cm. long, flowers singly inserted or in cymose groups; pedicels 1–5(–10 in fruit) mm. long, pubescent. Flowers whitish to yellow, sweet-scented; calyx divided ± halfway, lobes 2 mm. long, 1 mm. wide; petals rhomboid in outline, 1.5 mm. long, 1.5–2 mm. wide, scale bilobed, shaggy hairy. Stamens 6–8; filaments 4.5 mm. long, the basal part strikingly white hairy. Capsule reddish, (2–)3(–4)-gonous and -locular, 1.7–3 cm. long, 1.9–3 cm. wide, glabrescent, the angles acute or with wings up to 2 mm. broad; style and sepals persistent; endocarp pink with yellow margins, glabrous or nearly so. Seeds, dark brown or black, ovoid 1.5–2 cm. long, 6–8 mm. wide; aril bright yellow, up to 1 cm. long. Fig. 7.

UGANDA. Kigezi District: Kanungu, Oct. 1947, *Purseglove* 2542!; Busoga District: Butembe Bunya [Bunya], Nov. 1937, *Webb* 53!; Mengo District: Entebbe, Dec. 1934, *Chandler* 1188!

KENYA. Fort Hall District: 5 km. SW. of Muranga [Fort Hall], 9 Oct. 1956, *Dyson* 423!; Masai District: Masai Mara Game Reserve, Keekorok, Olchoro, June 1972, *Taiti* 2160!; Kwale District: Mrima Hill, 20 Nov. 1978, *Brenan et al.* 14629!

TANZANIA. Mwanza District: Ukerewe, Rubya Forest Reserve, 4 Oct. 1960, *Carmichael* 801!; Mbulu District: Lake Manyara National Park, Endabash [Ndabash] R., 18 Nov. 1963, *Greenway & Kirrika* 11041!; Songea District: 8 km. W. of Mtama, near Nyangao R., 14 Dec. 1955, *Milne-Redhead & Taylor* 7653!; Zanzibar, near Chukwani, 12 Feb. 1961, *Faulkner* 2754!

DISTR. **U** 2–4; **K** 3–7; **T** 1–4, 6–8; **Z**; **P**; tropical Africa from the west to east coast, south to Angola and South Africa

HAB. Evergreen forest, sometimes riverine or on forest margins, also in woodland and bushland, sometimes persisting in cultivations; 0–1900 m.

SYN. *B. zambesiaca* Baker in F.T.A. 1: 427 (1868). Type: Malawi, W. shore of Lake Nyasa, *Kirk* (K!, holo.)

Phialodiscus zambesiacus (Baker) Radlk. in Sitz. Akad. Wiss. München 9: 539 (1879) & in E. & P. Pf. III, 5: 345 (1895); P.O.A. C: 251 (1895); Z.A.E.: 481 (1912); Radlk. in E.P. IV, 165: 1148 (1933), F.P.N.A. 1: 523 (1948); T.T.C.L.: 560 (1949)

P. unijugatus (Baker) Radlk. in Sitz. Akad. Wiss. München 9: 539 (1879) & in E.P. IV, 165: 1147 (1933); I.T.U., ed. 2: 382 (1952)

P. plurijugatus Radlk. in Sitz. Akad. Wiss. München 20: 263 (1890); Hiern, Cat. Afr. Pl. Welw. 1: 171 (1896); Radlk. in E.P. IV, 165: 1148 (1933). Lectotype: Angola, Golungo Alto, *Welwitsch* 4519 (LISU, lecto., K!, BM, COI, G-DC, isolecto.)

P. verschuerenii De Wild. in B.J.B.B. 4: 361 (1914), as '*verschuereni*'; Radlk. in E.P. IV, 165: 1150 (1933). Type: Zaire, Congo da Lemba, *Verschueren* 742 (BR, holo.)

NOTE. Synott, on his 555 from Budongo Forest, states fruit up to 5 cm. but that seems to be unlikely judging by his specimen; probably due to a confusion with *B. welwitschii*. The fruits are used as a fish poison. *Bally* 8847 from Kilifi District, Gongoni Forest has dense disk-like galls 4 mm. diameter on the leaves.

2. **B. welwitschii** (*Hiern*) *Radlk.* in E.P. IV, 165: 1146 (1933); R. Wilczek in B.J.B.B. 21: 155, fig. 40 (1951); F.W.T.A., ed. 2, 1: 722 (1958); Hauman in F.C.B. 9: 315 (1960); Fouilloy & Hallé in Fl. Cameroun 16: 183 (1973) & in Fl. Gabon 23: 183, fig. 3, t. 43/1–7 (1973); Hamilton, Field Guide Uganda For. Trees: 223, fig. 415 (1981). Type: Angola, Golungo Alto, *Welwitsch* 4517 (LISU, holo., BM!, K!, iso.)

Dioecious tree 20–25(–35) m. tall, with grey bark showing small reticulate fissures or thin, smooth, hard and flaking; slash orange and yellow-brown mottled with scattered white pock marks; young twigs orange-pubescent but soon glabrescent and dark brown. Petiole up to 16 cm. long, flattened, narrowly winged; rhachis 4–15 cm. long; leaflets olive-green, in 2–4 pairs, narrowly elliptic to oblong-elliptic, the uppermost 8–40 cm. long, 3.5–10 cm. wide, with lower pairs becoming smaller with lowermost as little as half the size, shortly and bluntly acuminate at the apex, sometimes asymmetric at the base, glabrous apart from scattered hairs on the undersides of the veins; midrib impressed above, prominent beneath; lateral nerves in 7–10(–20) pairs, curving and sometimes joining inside the margin. Inflorescence-axis 3–10 cm. long; flowers in small groups or single; pedicels 5–8(–10 in fruit) mm. long. Sepals joined for $^1/_4$ of their length, lobes 2 mm. long, 1 mm. wide; petals rhomboid, 5 mm. long, 3 mm. wide, the scale equal to and joined to it to form a funnel-shaped compound petal. Stamens 7–8; filaments hairy in the lower part.

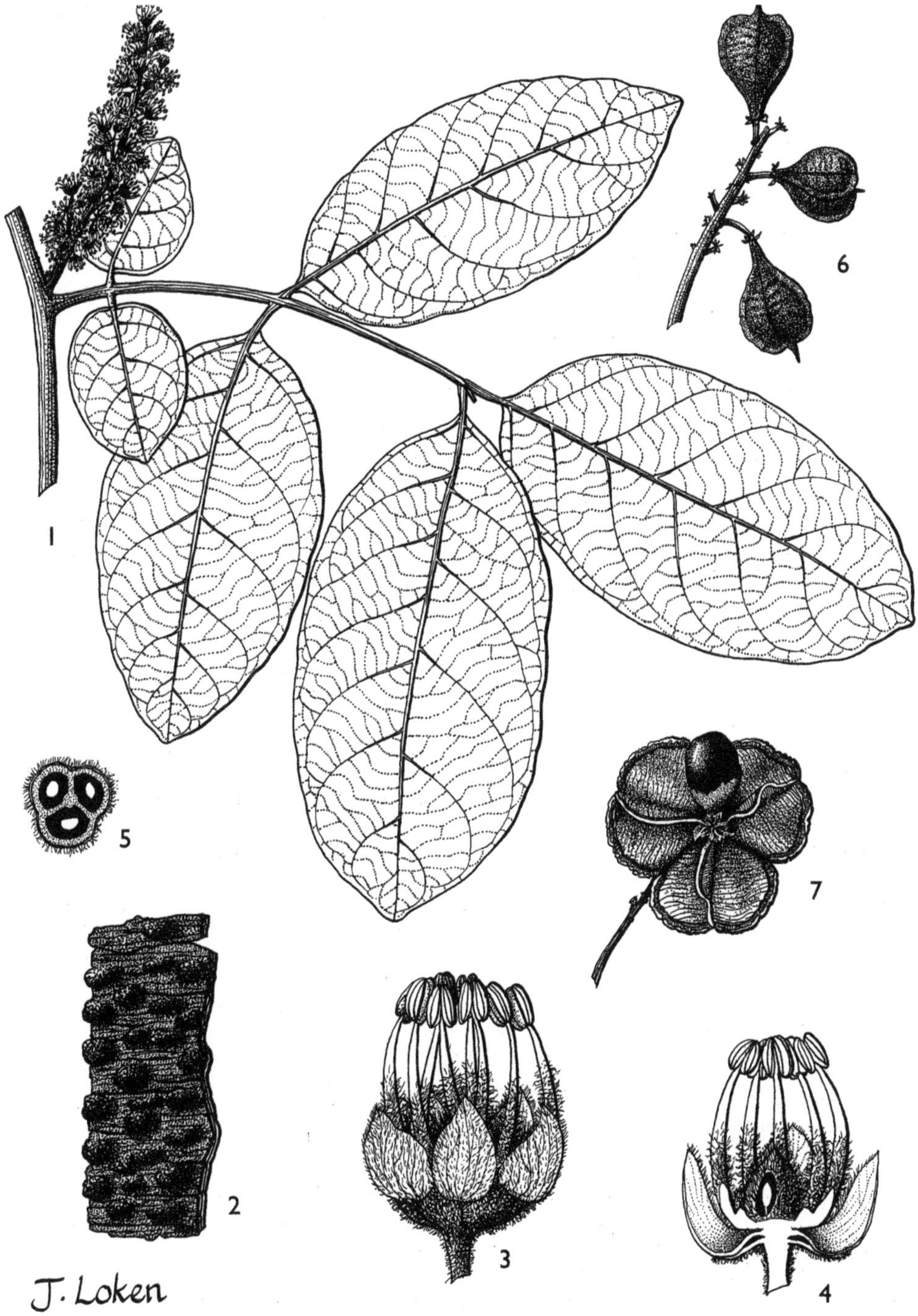

FIG. 7. *BLIGHIA UNIJUGATA* — **1**, leaf and inflorescence, $\times {}^{2}/_{3}$; **2**, bark showing warts, $\times {}^{2}/_{3}$; **3**, male flower, × 7; **4**, vertical section of flower, × 7; **5**, transverse section of ovary, × 7; **6**, infructescence, $\times {}^{2}/_{3}$; **7**, ripe fruit showing seed, $\times {}^{2}/_{3}$. 1, 3–5, from *Barbosa*; 2, from *Chase* 7400; 6, from *Torre* 3485; 7, from *Torre* 3901. Reproduced from F.Z. 2, t. 104 (1966).

Ovary 3-lobed; style entire, equalling the ovary. Capsule reddish or bright red, turbinate, 3–4-angled and -locular, up to 5.5–6(–8) cm. long, glabrous; endocarp creamy-velvety-pubescent. Seeds black or dark brown, 2.5 cm. long, 1.5 cm. wide; aril up to 2 cm. long.

UGANDA. Bunyoro District: Budongo Forest, Apr. 1935, *Eggeling* 1980! & same locality, plot 6, 18 May 1993, *Sheil* 1726!; Mengo District: Busiro, 1.5 km. N. of Kisubi on Mpigi road, 27 Mar. 1951, *Dawkins* 722!

DISTR. **U** 2, 4; Liberia, Nigeria, W. Cameroon, Zaire and Angola

HAB. Closed forest; 1050–1150 m.

SYN. *Phialodiscus welwitschii* Hiern, Cat. Afr. Pl. Welw. 1: 171 (1896)

Blighia laurentii De Wild. in Ann. Mus. Congo, Bot., sér. 5, 3: 113 (1909). Type: Zaire, Eala, *Laurent* 940 (BR, holo.)

B. mildbraedii Radlk. in Z.A.E.: 480, t. 64 (1912). Type: Zaire, Kivu, Beni, Muera, *Mildbraed* 2407 (B†, holo.)

Phialodiscus mortehanii De Wild. in B.J.B.B. 4: 361 (1914), as '*mortehani*'; Radlk. in E.P. IV, 165: 1149 (1933). Type: Zaire, Dundusana, *Mortehan* 966 (BR, holo.)

P. laurentii (De Wild.) Radlk. in E.P. IV, 165: 1149 (1933)

Blighia kamerunensis Radlk. in E.P. IV, 165: 1145 (1933); R. Wilczek in B.J.B.B. 21: 154 (1951). Type: Cameroon, Molundu, *Mildbraed* 4540 & near Lomie, between Assobam on the Bumba and Lomie, *Mildbraed* 5099 (B†, syn.)

B. wildemaniana Radlk. in E.P. IV, 165: 1145 (1933); I.T.U., ed. 2: 377 (1952). Type: Zaire, Kisantu, Stanley Pool, *Gillet* 1823 (BR, holo.)

NOTE. The seeds are used as a fish poison.

8. **ERIOCOELUM**

Hook.f., G.P. 1: 400 (1862); Radlk. in E.P. IV, 165: 1150 (1933)

Monoecious or dioecious trees or shrubs. Leaves paripinnate; leaflets in 2–5 pairs, petiolate or sometimes almost sessile with lowest pair ± stipuliform; other leaflets large, opposite or alternate, sometimes with translucent dots. Inflorescences paniculate or sometimes ± simply pseudoracemose, terminal or subterminal, always a few flowers of opposite sex in male and female inflorescences. Flowers regular; sepals 5; petals 5, 2–5 × as long as sepals; limb narrowly elliptic or lanceolate with a very short claw, scales much shorter than petals and often wider, not adnate, hairy; disk collar- or muff-shaped, funnel-like, often with sinuous margin. Stamens 8, ± as long as petals; filaments ± hairy or glabrous; staminodes 8, shorter than the petals. Ovary globose, 3-locular, each locule 1-ovulate; style thick, simple; rudimentary ovary reduced to a few hairs. Fruit capsular, ± globose, ± 3-angular, woody, loculicidally dehiscent; endocarp coloured, with tuft of fawn hairs at base of valves. Seeds erect, ovoid, smooth, with lobed orange or red arillode.

A genus of ± 10 species restricted to tropical Africa; only one occurs in the extreme north-west of the Flora area.

E. kerstingii *Engl.*, V.E. 3(2): 282, fig. 136 (1921)* & V.E. 1: 786, fig. 663 (1910), *nom. nud.*; F.W.T.A. 1: 502 (1928); Radlk. in E.P. IV, 165: 1153, fig. 34 (1933); F.P.S. 2: 340 (1952); I.T.U., ed. 2: 379 (1952); F.W.T.A., ed. 2, 1: 724 (1958); Hauman in F.C.B. 9: 324 (1960); Fouilloy & Hallé in Fl. Cameroun 16: 174, t. 41/1–5 (1973) & in Fl. Gabon 23: 174, t. 41/1–5 (1973); El Amin, Trees & Shrubs Sudan: 329 (1990). Types: none cited in original reference, but Togo, *Kersting* 87 (B) & Sokode-Basari, *Kersting* A518 (B, K!) must qualify as syntypes**

* This is barely adequately published; those disagreeing will have to date the name from Hutchinson & Dalziel (1928).

** Togo, Sokode, *Kersting* A257 (K!) is presumably another although not cited by Fouilloy & Hallé.

FIG. 8. *ERIOCOELUM KERSTINGII* — **1**, flowering branchlet, × $^2/_3$; **2**, young male flower, × 6; **3**, mature male flower, × 6; **4**, male flower with petals removed, × 6; **5**, female flower, × 6; **6**, female flower with sepals turned back and cut, × 6; **7**, fruits, × 1; **8**, seed, × 2. 1–4, from *Hepper* 1518; 5, 6, from *Silitoe* 313; 7, 8, from *Louis, Breteler & de Bruijn* 388. Drawn by Pat Halliday.

Tree or shrub 8–18(–30) tall; bark grey-brown; foliage appearing dark; young stems, petioles, inflorescences etc. covered with dense ferruginous or golden brown hairs. Leaves with 2–3 pairs of opposite or subopposite leaflets, the lowest pair stipuliform, often soon deciduous but leaving scars; leaflet-blades oblong, oblong-elliptic, obovate or oblanceolate, 2–20 cm. long, 1.5–8.6 cm. wide, subacute, obtuse or rounded at the apex, cuneate and often unequal or ± rounded at the base, ± glabrous above, ± pilose or pubescent on the venation beneath or eventually glabrous. Panicles large, ± divided from the base, 8–12(–20) cm. long, 10–15 cm. wide, with primary branches up to 8 cm. long; pedicels 4 mm. long; all axes densely ferruginous pubescent. Flowers ± 4 mm. wide; sepals oblong, 1.5 mm. long, 0.5 mm. wide, pubescent on both faces; petals white, linear-lanceolate, 2.5 mm. long, scales very hairy, scarcely as wide as the limb; disk 1 mm. tall. Stamens 4 mm. long; staminodes 2 mm. long. Ovary globose, 2.5 mm. diameter, densely ferruginous pubescent; style 3 mm. long; rudimentary ovary reduced to few hairs. Fruit light brown, yellow or red, depressed globose, ± woody, 1.8–3 cm. diameter, at first hairy, eventually smooth and glabrous; valves 2–2.5 cm. wide, 2.5–3 mm. thick, woolly at base inside. Seeds 3*, black, 1.25 cm. long, shiny with red arillode. Fig. 8.

UGANDA. W. Nile District: Madi, Butze, *Dawkins* 330 (*fide* I.T.U.)
DISTR. U 1; Mali, Guinea Bissau, Sierra Leone, Liberia, Ivory Coast, Ghana, Togo, Benin, N. Nigeria, Cameroon, Gabon, Central African Republic, Zaire and Sudan
HAB. Riverine forest; ± 800 m.

NOTE. Hauman described a var. *katangense* with more coriaceous elliptic leaflets with a more distinct reticulation and petal scales distinctly broader than the limb. The Kew isotype has distinctly subcordate leaves and I think the taxon is best referred to *E. lawtonii* Exell. Much West African material referred to *E. kerstingii* has distinctly acute or acuminate leaves.

9. HAPLOCOELOPSIS

F.G. Davies in K.B. 52: 231–234 (1997)

Shrub or small tree with dense leafy crown; probably monoecious. Leaves pinnate, leaflets entire; lowermost leaflets reduced, stipule-like, one of the pair inserted on or shortly above the petiole, sometimes caducous. Inflorescence axillary, racemose, slender, simple or branched. Flowers regular; sepals 5, united for a very short distance above the base, the lobes unequal, oblong-ovate; petals 5, elliptic-lanceolate to ovate, exceeding the sepals at maturity, scale bilobed, half as large as the limb; disk a complete ring, flat, broad, glabrous. Male flower: stamens 9, glabrous. Female flower: staminodes 8, curved in bud, glabrous; ovary 2-lobed, 2-locular; style truncate, with two stigmatic grooves. Fruit subglobose, grooved.

H. africana *F.G. Davies* in K.B. 52: 232, fig. 1 (1997); Beentje in Utafiti 1(3): 111, fig. 8, map (1988) & K.T.S.L.: 418, fig. (1994), *nom. invalid.*, descript. anglice. Type: Kenya, Lamu District, Witu, Gongoni Forest, *Dale* in *F.H.* 3820 (BM!, holo., EA!, K!, iso.)

Evergreen shrub or small tree to 10 m. tall; branches slender, grey-brown, initially coarsely hairy. Petiole 0–0.5 cm. long; rhachis 3–8 cm. long, with white or brown bristly hairs; petiolules 0–5 mm. long, fleshy, moderately or sparsely hairy; lowermost leaflet reniform or ovate, 0.5–2 cm. long, 0.4–2.5 cm. wide, the second leaflet of the pair often inserted further up the rhachis and somewhat larger; uppermost leaflets rhomboid-elliptic or lanceolate, acute or rounded at the extreme apex and almost acuminate, 5–16 cm. long, 2–7 cm. wide; blade olive-green, glabrous apart from the

* El Amin mentions 3–4 seeds so fruit may rarely be 4-locular.

FIG. 9. *HAPLOCOELOPSIS AFRICANA* — **1**, flowering branchlet, × $^2/_3$; **2**, male flower (with anthers fallen), × 8; **3**, female flower, × 8; **4**, female flower with petals removed, × 8; **5**, petal, × 8; **6**, fruit, × 3. 1–5, from *Dale* 3820; 6, from *Luke & S.A. Robertson* 247 & *Mwasumbi* 12523. Drawn by Margaret Tebbs. Reproduced from K.B. 52, fig. 1 on p. 233 (1997).

midrib; lateral veins faint, in 7–10 pairs. Inflorescence 3.5–11 cm. long, with up to 2 branches; rhachis and pedicels pilose; flowers in clusters of 2–3, subtended by caducous bracts 0.5 mm. long; pedicels 1.6–2 mm. long. Calyx-lobes oblong-ovate, the 2–3 larger ones 1.5–2 mm. long, 1–1.2 mm. wide, externally pubescent (very sparsely so in Angola) and slightly pubescent inside at the top. Male flowers: petals narrowly elliptic-oblong, 2 mm. long, 1.5 mm. wide; filaments 1.3 mm. long, glabrous; rudimentary ovary 0.5 mm. long. Female flower: petals narrowly elliptic-lanceolate to ovate, 2–3 mm. long, 1–1.5 mm. wide, glabrous, ± venose, with an obtriangular slightly thickened claw 0.8 mm. long and inflexed scale with ciliate and erose margin; staminodes 2 mm. long; ovary ovoid, 2 mm. long and wide, pubescent or almost glabrous; style ± 1 mm. long. Fruit yellow, subglobose, 2.5 cm. wide, grooved between the seeds; pedicels to 6 mm. long. Fig. 9.

KENYA. Kwale District: Shimba Hills, Makadara Forest, 18 Feb. 1987, *Luke & S.A. Robertson* 247! & Jombo [Dzombo] Hill, 7 Feb. 1989, *Mrima–Dzombo Exped.* 142!; Lamu District: Ras Tenewi, *Luke & S.A. Robertson* 1433!

TANZANIA. Uzaramo District: Mfyoza, Banda Forest Reserve, 12 Nov. 1969, *Ruffo* 304!; Newala District: Makonde thicket, Nov. 1953, *Eggeling* 6733! & Newala, 12 Dec. 1943, *Gillman* 1070!

DISTR. **K** 7, **T** 6, 8; Angola (see note)

HAB. Lowland wet and dry forest and thickets; 40–700 m.

SYN. *Haplocoelum sp.* sensu T.T.C.L.: 558 (1949) quoad *Gillman* 1070
Melanodiscus sp. nov. sensu K.T.S.: 515 (1961) quoad *Dale* in *F.H.* 3820

NOTE. The affinities of this genus are with *Eriocoelum* and *Aporrhiza* in the Cupanieae, a tribe where the common form of fruit is a capsule dehiscing usually into 3 (rarely 2) valves to reveal seeds with brightly coloured, cup-shaped arils around their proximal ends.

Leaf size appears to decrease from **K** 7 to **T** 8, but the three collections from Newala may all derive from the same individual tree so no great significance should be attached to its smaller leaves for the time being. The Angolan plant is almost glabrous, with the sepals hairy only near the apex, but otherwise similar vegetatively; the flowers are immature.

10. STADMANIA

Lam., Tab. Encycl. 2: 443, t. 312 (1793); Radlk. in E.P. IV, 165: 1009 (1933), as '*Stadmannia*'; Capuron in Mém. Mus. Nat. Hist. Nat., Paris, 19: 151 (1969)

Pseudolitchi Danguy & Choux in Bull. Mus. Hist. Nat. Paris 32: 390 (1926); Radlk. in E.P. IV, 165: 1498 (1933)

Tree. Leaves pinnate; leaflets entire or serrate. Inflorescence an axillary thyrse bearing clusters of yellowish flowers, male and female in the same inflorescence or (not in mainland Africa) the trees dioecious. Calyx opening precociously, united in a cup to at least half-way, shallowly 5-lobed; petals absent or (not in mainland Africa) 5; disk cushion-like, hairy. Stamens (6–)8. Ovary 3-lobed, 3-locular with one ovule per locule; style equal to the ovary, 3-furrowed. Fruit of 1–3 spherical mericarps, dehiscing along a vertical line. Seeds spherical, nearly totally covered in a fleshy aril.

One species, with 2 subspecies, in mainland Africa, Madagascar and the Mascarenes, plus 5 endemic species in Madagascar.

S. oppositifolia *Poir.* in Lam., Encycl. Méth. Bot. 7: 376 (1806); Beentje, K.T.S.L.: 423, fig. (1994); F. Friedmann in Fl. Masc. 76: 16, t. 6–9 (1997). Type: Mauritius [Isle de France], *Stadman* 419 (P-LA, holo., photo.!)

Medium-sized tree 9–30 m. tall; bark yellow or brown, smooth, finely fissured or concentrically-flaking, mottled yellow and grey, slash pink; twigs initially rusty- or tawny-pubescent. Petiole 3–7 cm. long, angled, pubescent; rhachis 4–10 cm. long;

FIG. 10. *STADMANIA OPPOSITIFOLIA* var. *OPPOSITIFOLIA* — **1**, flowering branchlet, × $^{2}/_{3}$; **2**, male flower, × 6; **3**, **4**, female flower, × 6; **5**, seed, × 1. 1, 5, from *S.A. Robertson* 5658; 2, from *Bradburne* 102; 3, 4, from *Gillet & S.A. Robertson* 24025. Drawn by Pat Halliday.

petiolules 1–7 mm. long; leaflets in (1–)2–4 pairs, opposite, the uppermost pair largest, blade narrowly elliptic or elliptic, 4–15 cm. long, 2.5–6.5 cm. wide, emarginate or rounded at the apex; asymmetrically cuneate at the base, entire, thickened, thick-textured, glabrous apart from the midrib, microscopically glandular on both surfaces, shining above, matt beneath; lateral veins in 10–16 pairs, indistinct, venation finely reticulate. Inflorescences 4–12.5 cm. long; flowers yellow in cymose clusters, scented; pedicels 2–3 mm. long, increasing to 5 mm. in fruit. Calyx cup 0.5 mm. long; lobes broadly triangular, 0.2 mm. long, densely pubescent. Stamens 8; filaments in the bud straight and very short, at maturity 3 mm. long, pilose or glabrous; anthers oblong, pilose or glabrous. Style shortly 3-fid. Fruit golden yellow, usually of one developed globose mericarp, 1–2.1 cm. in diameter, shortly velvety pubescent, with the abortive carpels remaining attached to the base. Seed chestnut brown, globose or ovoid, 1–1.2 cm. in diameter; aril red, loosely covering seed. Fig. 10.

subsp. **oppositifolia**

Leaves larger, pubescence of ± adpressed hairs; leaflets with obscure reticulation beneath. Inflorescences longer.

var. **oppositifolia**; Capuron in Mém. Mus. Nat. Hist. Nat., Paris, 19: 154, t. 40/1–2 (1959)

Leaflets larger, 9–15 cm. long, 3.5–6 cm. wide. Inflorescences longer. Staminal filaments ciliate.

KENYA. Kwale District: Vanga, Kirao, Jan. 1930, *R.M. Graham* in *F.H.* 2232! & S. Diani Forest, near sea, Coverdale Plot, 1 Jan. 1983, *Gillett & S.A. Robertson* 24025! & 24028! & Kinondo (Ngalani Kaya) Forest, 14 Feb. 1977, *R.B. & A.J. Faden* 77/387!
TANZANIA. Pangani District: Pangani R., 28 Sept. 1966, *Bradburne* 102!; Rufiji District: Mafia I., Kanga–Miskitini, 15 Aug. 1937, *Greenway* 5112!; Zanzibar, Chuini, 6 July 1933, *Vaughan* 2131!
DISTR. **K** 7, **T** 3, 6; **Z**; Madagascar and Mauritius, formerly in Réunion (extinct)*
HAB. Dry evergreen forest and coastal bushland on coral rag; 0–5 m.**

SYN. *S. sideroxylon* DC., Prodr. 1: 615 (1824); Radlk. in E.P. IV, 165: 1010 (1933); K.T.S.: 516 (1961), *nom. illegit.* Type as for *S. oppositifolia*

NOTE. A second variety of subspecies *oppositifolia*, var. *grevei* (Danguy & Choux) Capuron in Mém. Mus. Nat. Hist. Nat., Paris, 19: 154 (1959), based on *Pseudolitchi grevei* Danguy & Choux, is confined to Madagascar. Subsp. *rhodesica* Exell from Zimbabwe has smaller leaflets, 4–10 × 2–4 cm., shorter petiolules and a more pronounced vein reticulum on the leaflet underside.
The superfluous epithet *sideroxylon* was published because of objection to the implication of opposite leaves in '*oppositifolia*' (it is the leaflets that are opposite) but this is not a legitimate reason for rejecting the epithet.
Despite the number of districts with records of this subspecies it is apparently rather rare. The wood is very hard, known locally as 'iron-wood', and is used for construction.

11. **PAPPEA**

Eckl. & Zeyh., Enum. Pl. Afr. Austr.: 53 (1834); Radlk. in E.P. IV, 165: 1012 (1933)

Evergreen monoecious trees or shrubs. Leaves simple, petiolate, in clusters at the tips of branches. Inflorescence a lax or dense thyrse of yellow flowers. Calyx opening precociously to reveal the anthers, in male flowers cup-like, barely 5-lobed, in female flowers lobed to halfway. Petals minute (?sometimes absent). Stamens 8(–10);

* Derived from an ambiguous statement of Rivals in Trav. Lab. For. Toulouse, Tom. 5, Vol. 1, Art. 2: 152 (1952), but it is not mentioned in Cordemoy.
** Davies gave the altitude limit as 1000 m., but this must refer to extra East African material. [B.V.]

filaments very short in bud. Ovary 3-lobed; stigma 3-furrowed. Fruit a spherical mericarp, containing a single seed, dehiscing along a vertical suture into 2 valves; all 3 mericarps developed or if not then 1–2 abortive at base of fertile one. Seed ovoid to spherical; aril red, fleshy, enclosing the seed.

One species, varying greatly in leaf form, from Zaire to Somalia and south to the Cape.

P. capensis *Eckl. & Zeyh.*, Enum. Pl. Afr. Austr.: 53 (1834); Radlk. in E.P. IV, 165: 1013, fig. 26 (1933); K.T.S.: 516, fig. 94 (1961); F.F.N.R.: 225 (1962); Exell in F.Z. 2: 540, t. 116 (1966); Troupin, Fl. Rwanda 2: 308, fig. 96/1 (1983); Blundell, Wild Fl. E. Afr.: 137, t. 347 (1987); Noad & Birnie, Trees of Kenya: 265, fig. on 266 (1989); Vollesen in Fl. Ethiopia 3: 508, fig. 125.13 (1990); Beentje, K.T.S.L.: 422, fig., map (1994). Lectotype, chosen here: South Africa, W. Cape Province, Fish R., *Zeyher* 151 (K!, lecto.)

Much-branched shrub or tree to 15 m. tall; crown thick and rounded; bark dark grey, reticulate, rough or smooth save for small tesselations, variable according to fire regime; twigs tawny-pubescent when young, knobbly with raised leaf-scars, soon glabrescent, dark grey. Petiole 0.3–1.7 cm. long, pubescent; petioles 0.7–1.8 cm. long; leaf-blades yellow-green, oblong or broadly or narrowly elliptic or obovate or occasionally oblong-lanceolate, 3–18 cm. long, 1.3–8 cm. wide, apex obtuse or emarginate or shortly acuminate, rarely acute, base asymmetric, truncate or rounded or cordate on one side, entire and undulate or rather rarely sharply serrate, the teeth 1–2.5 mm. deep, glabrous above apart from hairs on the larger veins, the lower surface softly hairy and glandular; lateral nerves in 12–20 pairs, strongly branched, venation closely reticulate. Inflorescences spicate, in leaf-axils on short lateral shoots; axes 3–15(–25) cm. long, with dense subsessile clusters of white, yellow-green or yellow flowers; pedicels 1–5 mm. long. Calyx 0.6–0.8 mm. long, densely whitish pubescent; petals 5, claw 0.6 mm. long, limb triangular, 0.4 mm. long, long-white ciliate, scale 0.2 mm. long; disk said to be glandular. Stamens much exserted; filaments eventually 4 mm. long, pilose in the lower part; anthers triangular-ovate, with scattered hairs, the apex of the connective inflated; staminodes with 2 mm. long filaments and slightly smaller anthers. Ovary rudiment a small cushion of hairs. Fruit green-velvety when young, brown when mature, spherical, 8–12 mm. diameter, with a longitudinal ridge along which dehiscence occurs; interior of the valves glabrous. Seed 6–9 mm. across. Fig. 11.

UGANDA. Acholi District: Paimol, Dec. 1935, *Eggeling* 2346! & 2358!; Ankole District: Sanga Rest Camp, Oct. 1932, *Eggeling* 612!; Masaka District: W. Masaka hills, 64 km. on Mbarara road, Oct. 1952, *Philip* 521!

KENYA. W. Suk District: 19 km. N. of Kapenguria on Lodwar road, 14 Sept. 1970, *Aggundey* 5!; Machakos District: Lukenya Hill, 22 Aug. 1953, *Bally* 9055!; Masai District: W. Ngong Hills, 3 Sept. 1965, *Kokwaro, Kanuri & Mathenge* 313!

TANZANIA. Musoma District: Mugango, 2 June 1959, *Tanner* 4305!; Masai District: Loliondo, near Iyasi, Mulubani, 5 July 1956, *G.R. Williams* 683!; Songea District: S. of Songea, Matagoro Hills, 22 Feb. 1956, *Milne-Redhead & Taylor* 8862!

DISTR. **U** 1, 2, 4; **K** 1–7; **T** 1–3, 5, 7, 8; Zaire, Rwanda, Ethiopia, N. Somalia, south to South Africa

HAB. Grassland, often with scattered trees, woodland and scrub, often in rocky places; 900–2000 m.

SYN. *P. schumanniana* Schinz in Abh. Bot. Ver. Prov. Brandenb. 30: 156 (1888). Type: Namibia, Great Namaqualand, *Schenck & Schinz* (Z, holo.)

P. radlkoferi Schweinf. in Bull. Herb. Boiss. 7, app. 2: 338 (1899); Penzig in Atti. Congr. Bot. Int.: 27 (1892), *nom. semi-nud.*; Radlk. in E.P. IV, 165: 1015 (1933). Types: Ethiopia, Gheleb, *Schweinfurth & Riva* 1068 (ubi?, syn.) & 1080 (ubi?, syn.) & 1123 (ubi?, syn.) & Acrour, *Schweinfurth & Riva* 1041 (FT!, K!, P!, isosyn.) & near Aidereso, Galata, *Schweinfurth & Riva* 1617 (FT!, K!, isosyn.)

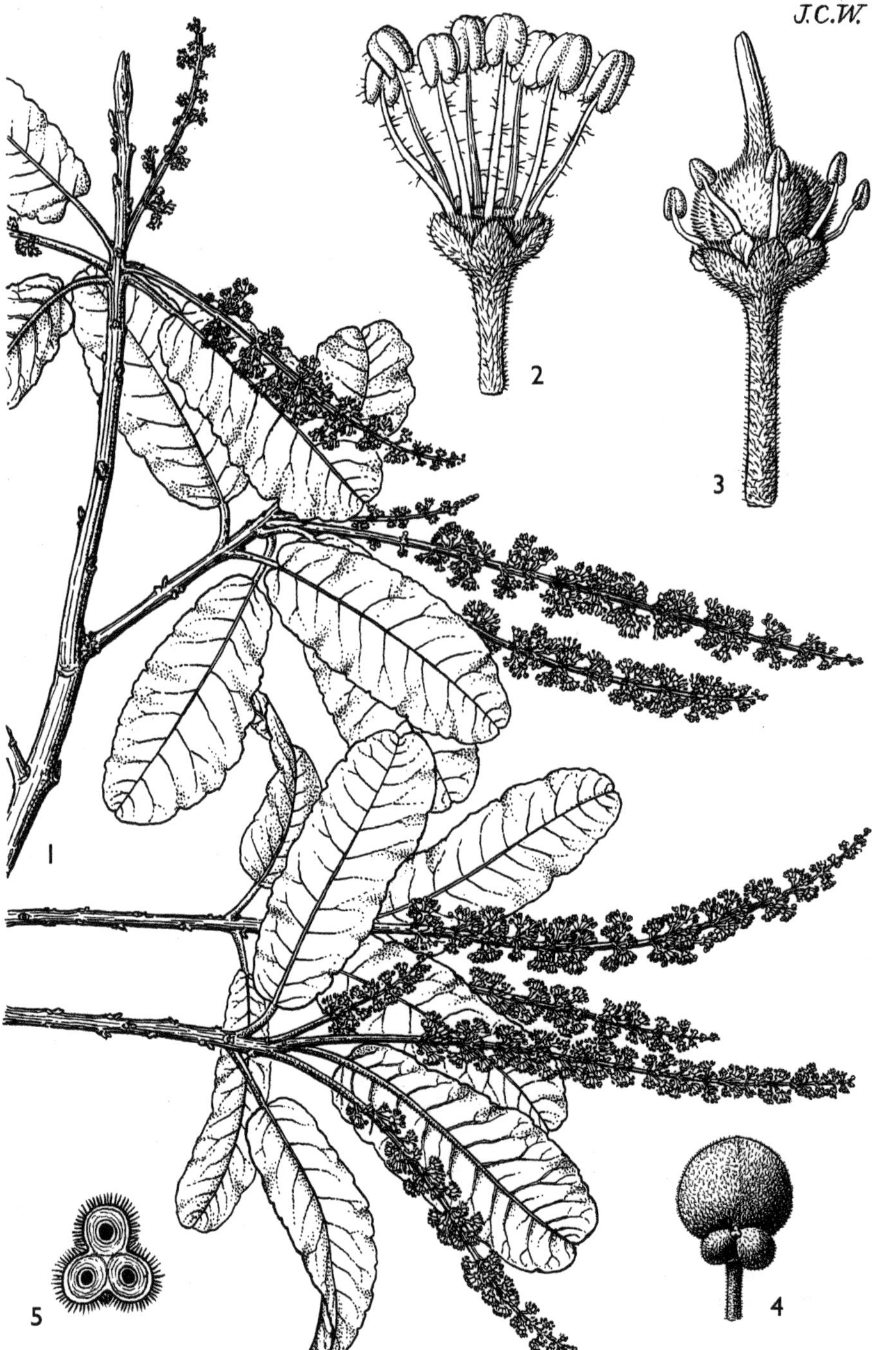

FIG. 11. *PAPPEA CAPENSIS* – **1**, flowering branch, × $^{2}/_{3}$; **2**, male flower, × 8; **3**, female flower, × 8; **4**, fruit, × 2; **5**, transverse section of fruit, × 2. 1, 2, from *Chase* 633; 3, from *Fanshawe* 5724; 4, 5, from *Brenan* 7764. Reproduced from F.Z. 2, t. 116 (1966).

P. ugandensis Baker f. in J.L.S. 37: 138 (1905); Radlk. in E.P. IV, 165: 1014 (1933); T.T.C.L.: 560 (1949); I.T.U., ed. 2: 382, fig. 76 (1952); Hauman in F.C.B. 9: 381 (1960). Type: Uganda, Ankole District, Mulema, *Bagshawe* 369 (BM!, holo.)

P. capensis Eckl. & Zeyh. var. *radlkoferi* (Schweinf.) Schinz in Viert. Nat. Ges. Zürich 53: 490 (1908)

P. fulva Conrath in K.B. 1908: 221 (1908). Type: South Africa, Gauteng [Transvaal], Modderfontein, *Conrath* 295 (K!, holo.)

P. radlkoferi Schweinf. var. *angolensis* Schltr. in E.J. 50, suppl.: 422 (1914); C.F.A. 2: 91 (1954). Type: Angola, Gambos, *Antunes* 184 (B†, holo., ?P, iso.)

P. radlkoferi Schweinf. var. *ugandensis* (Baker f.) Schltr. in E.J. 50, suppl.: 422 (1914)

NOTE. Working at Kew in 1948 I hesitated to accept that *P. ugandensis* was entirely synonymous with *P. capensis* which has on average much smaller leaves; it is possible that subspecies could be recognised. True *ugandensis* extends to the Transvaal. [B.V.]

The red fleshy aril is edible but the seed contains a somewhat purgative oil. The leaf-margin is occasionally sharply subspinous-serrate both in the Cape and more northern populations. The crushed bark makes a bright red drink used by the Masai.

12. MACPHERSONIA

Blume, Rumphia 3: 156 (1847); Radlk. in E.P. IV, 165: 888 (1932); Capuron in Mém. Mus. Nat. Hist. Nat., Paris, 19: 117 (1969)

Small trees or shrubs, dioecious in the Flora area. Leaves pinnate (not in East Africa) or bipinnate; leaflets entire. Inflorescence a paniculate or unbranched thyrse. Flowers regular; sepals 5, free, imbricate; petals 5, small, clawed, infolded at the lateral margins to produce 2 scales; disk circular, flat, barely lobed. Stamens 7–8; filaments folded in bud; anthers sagittate at the base; staminodes similar with shorter filaments. Ovary 3-locular, spherical; style 3-furrowed. Fruit a spherical capsule without exterior dehiscence lines, splitting into 3 valves (but material from our area indehiscent or only partially splitting at apex). Seeds 1–3 per capsule, each largely covered by a fleshy aril.

Six species in Madagascar and neighbouring islands, with one species extending to eastern Africa.

M. gracilis *O. Hoffm.*, Sert., Pl. Madag.: 14 (1881); Radlk. in E.P. IV, 165: 892 (1932). Type: Madagascar, Nosy Be, *Hildebrandt* 3124 (B†, holo., K!, BM!, iso.)

Small tree or shrub to 10 m. tall. Young axes densely golden-hairy with crisped short hairs and long spreading coarse hairs; older twigs glabrous, grey or black. Leaves bipinnate, superficially legume-like; petiole 0.3–0.8 cm. long; primary rhachis (4–)6–20 cm. long, pubescent; secondary rhachides in 4–10 pairs, mostly opposite, pubescent; leaflets in (3–)8–18 pairs, sessile, the lowermost shorter, otherwise oblong, 1–2.5 cm. long, 2.5–8 mm. wide, apex rounded or slightly emarginate or very minutely apiculate or (not in East Africa) finely acuminate, the base obliquely wedge-shaped, glabrous and glossy apart from hairs on the margin and midrib; midrib conspicuous, lateral nerves fern-like, closely-set and not prominent. Thyrse unbranched, 4–13(–20) cm. long; flowers single and in groups of 2–4; pedicels of male flowers 1 mm., of female flowers 2 mm. long; subtending bracts 0.2–0.5 mm. long. Sepals broadly ovate, concave, unequal, 1.3–2 mm. long, thin-textured, glabrous apart from ciliate margin; petals broadly triangular, 1–1.5 mm. long, 1.5–1.8 mm. wide, toothed, the lateral margins infolded, very densely hairy. Stamens 7–8, exserted; anthers 0.8 mm. long, glabrous. Ovary rudiment in male flowers very reduced. Capsule red or purple, 1.1–1.5 cm. wide, glabrous. Seed single.

FIG. 12. *MACPHERSONIA GRACILIS* var. *HILDEBRANDTII* — **1**, flowering branchlet, × 1; **2**, leaflet, × 2; **3**, tip of young inflorescence, × 4; **4**, male flower, × 6; **5**, **6**, sepals from male flower, × 6; **7**, petal from male flower, × 12; **8**, stamen from male flower, × 6; **9**, female flower, × 6; **10**, sepal from female flower, × 6; **11**, petal from female flower, × 12; **12**, female flower with sepals removed, × 6; **13**, staminode from female flower, × 6; **14**, infructescence, × 1. All from *Vaughan* 1211. Drawn by D.R. Thompson.

var. **hildebrandtii** (*O.Hoffm.*) *Capuron* in Mém. Mus. Nat. Hist. Nat., Paris, 19: 122 (1969); Beentje, K.T.S.L.: 421, fig. (1994). Type: Madagascar, Nosy Komba, *Hildebrandt* 3245 (B†, holo., BM!, K!, iso.)

Apex of leaflets rounded, slightly emarginate or very minutely apiculate. Fig. 12.

KENYA. Kwale District: near Shimoni, 19 Feb. 1976, *Wamukoya* x41!; Kilifi District: Mangea Hill, 25 Feb. 1987, *Luke & S.A. Robertson* 265!

TANZANIA. Pangani District: Mwera, Mruazi R., Dec. 1964, *Procter* 2785!; Uzaramo District: Dar es Salaam, Oyster Bay, 6 Mar. 1973, *B.J. Harris & Mwasumbi* 6777!; Rufiji District: Mafia I., N. of Bweni, 12 Aug. 1937, *Greenway* 5071!; Zanzibar, Chaani–Pwani Mchangani, 1 Dec. 1930, *Greenway* 2612!; Pemba I., Chake Chake Creek, 24 Dec. 1930, *Greenway* 2783!

DISTR. **K** 7; **T** 3, 6, 8; **Z**; **P**; Mozambique, Madagascar, Comoro Is. and Aldabra

HAB. Coastal bushland, thicket and forest edges, often on coral outcrops or along streams, salt tolerant; 0–100 m.

SYN. *M. hildebrandtii* O. Hoffm., Sert. Pl. Madag.: 14 (1881); T.T.C.L.: 559 (1949); Exell in F.Z. 2: 534, t. 113 (1966); Radlk. in E.P. IV, 165: 891 (1932)
[*M. madagascariensis* sensu P.O.A. C: 251 (1895), *non* Blume (1847)]

NOTE. *M. gracilis* var. *gracilis* found in Madagascar differs in having finely acuminate leaflets.

13. LECANIODISCUS

Benth. in Hook., Niger Fl.: 250 (1849); Radlk. in E.P. IV, 165: 879 (1932)

Chiarinia Chiov., Result. Sci. Miss. Stef.-Paoli, Coll. Bot.: 149 (1916)

Small or medium-sized dioecious trees. Leaves pinnate; leaflets entire. Inflorescences short or long axillary racemoid thyrses. Flowers regular; sepals 3–5, concave, free or partly joined; petals absent (or 5 in West Africa); disk lobed. Stamens 8–12; filaments glabrous, folded in bud. Ovary unlobed, 3-locular; stigma subcapitate, subsessile, with 3 lobes. Fruit a slightly asymmetrically ellipsoid capsule, dehiscing by an irregular split, or indehiscent. Seed usually single by abortion of 2 ovules, enclosed in an aril.

A genus of 3 species in tropical Africa, one being rather variable in leaf form.

Leaf-rhachis 18–26 cm. long; fruit-hairs drying orange-ferruginous 1. *L. cupanioides*
Leaf-rhachis 2–15 cm. long; fruit-hairs drying greyish · · · · · · · · 2. *L. fraxinifolius*

1. **L. cupanioides** Benth. in Hook., Niger Fl.: 251 (1849); F.T.A. 1: 429 (1868); Radlk. in E.P. IV, 165: 880 (1932); I.T.U., ed. 2: 379 (1952); F.P.S. 2: 341 (1952); F.W.T.A., ed. 2, 1: 720 (1958); Hauman in F.C.B. 9: 368 (1960); Fouilloy & Hallé in Fl. Cameroun 16: 143, t. 33/1–6 (1973) & in Fl. Gabon 23: 143, t. 33/1–6 (1973); Hamilton, Field Guide Uganda For. Trees: 223 (1981). Lectotype, chosen here: Ghana, Accra, *Vogel* (K!, lecto.)

Small tree ± 12 m. tall, with spreading crown; branches dull brown; twigs initially pubescent and glandular. Petiole 4–12 cm. long, flattened; rhachis 18–26 cm. long, puberulent; petiolules 2–8 mm. long; leaflets in 5–7 pairs, ovate-elliptic, elliptic or oblanceolate, the lowermost relatively broader in relation to its length, the middle and upper leaflets 9–18 cm. long, 3–9 cm. wide, shortly acuminate, the base obtuse, glabrous; lateral nerves in 10–15 pairs. Inflorescence-axis 7–11 cm. long, all parts initially golden-pubescent. Male flowers: pedicels 5–6 mm. long; calyx ovoid-spherical in bud, 3.2 mm. long, the lobes very short, then opening nearly to the base and reflexed; disk conspicuous, fleshy; stamens exserted. Female flowers: pedicels 2–3 mm. long, not accrescent in fruit; calyx ovate-oblong in bud, as in the male, scarcely visibly lobed at this stage, later opening fully and reflexed; disk obscured by the pinkish- or brownish-pubescent ovary; staminodes rudimentary, the anthers degenerate. Fruit orange when fresh, drying orange-ferruginous, asymmetrically ovoid, 2–2.2 cm. long, 1.5–1.7 cm. wide, indumentum velvety; seed slightly compressed, 1.1 cm. long, 1.2 mm. wide.

UGANDA. Toro District: Bwamba, Kabango, 21 Nov. 1935, *A.S. Thomas* 1489! & Bwamba, Feb. 1944, *Mukasa* in *E.A.* 5527!

DISTR. **U** 2; Senegal to Sudan, south to Zaire and Angola

HAB. Lowland rain-forest, periodically inundated forest and riverine fringes; 900 m. (see note)

NOTE. Fouilloy & Hallé include Kenya in the distribution given in Fl. Cameroun but no material has been seen from there. The bark has been used as a purgative in Uganda. Davies gave a much lower altitude limit but presumably referring to West Africa. [B.V.]

2. **L. fraxinifolius** *Baker* in F.T.A. 1: 429 (1868), as '*fraxinifolia*'; Radlk. in E.P. IV, 165: 881 (1932); T.T.C.L.: 558 (1949); K.T.S.: 514 (1961); F.F.N.R.: 225 (1962); Exell in F.Z. 2: 530, t. 111 (1966); Hamilton, Field Guide Uganda For. Trees: 223 (1981); Friis in K.B. 39: 781 (1984); Vollesen in Fl. Ethiopia 3: 504 (1990); Beentje, K.T.S.L.: 420, map & fig. (1994). Types: Mozambique, Sena, *Kirk* & Malawi, Shire R., *Kirk* (both K!, syn.)

Untidy tree to 20(–30) m. tall with pale grey bark, sometimes slightly buttressed; young twigs, petioles, rhachides and petiolules tawny-pubescent and sometimes with sessile glands. Leaves often pink when very young; petiole 1–6 cm. long, rhachis 2–15 cm. long; petiolules 0–3(–5) mm. long; leaflets in 2–6 pairs, opposite or alternate, very varied in shape (see subspecies below), lanceolate, elliptic, ovate or obovate, the uppermost pair the largest, 4–12 cm. long, 2–3.5 cm. wide, variously long tapering to rounded at the apex, often slightly asymmetric at the base, undulate in dried specimens, glabrescent or sparsely pilose, with or without sessile glands; lateral nerves in 5–16 pairs, prominent beneath. Inflorescences axillary on the main stems and in the axils of immature leaves on short shoots; axes (1–)4–12 cm. long, with or without lateral branches; pedicels 1–3(–5) mm. long, densely pubescent. Sepals 4, 2.5–3 mm. long, pubescent outside; disk ciliate. Stamens 8–11, exserted; staminodes with shorter filaments. Ovary flask-shaped, 2–3 mm. long, densely tomentose; style truncate, persisting on the fruit. Fruit pink when fresh, greyish tomentose when dry, 1–1.5 cm. long, said to be blue internally. Seed ellipsoid, 1–1.2 cm. long, 8 mm. wide.

subsp. **vaughanii** (*Dunkley*) *Friis* in K.B. 39: 781, map 1 (1984); Vollesen in Fl. Ethiopia 3: 505 (1990). Type: Zanzibar, Chwaka, *Vaughan* 2301 (K!, holo., EA, iso.)

Tree or shrub 4–21(–?24) m. tall. Leaflets dark or dull green, lanceolate or ovate-elliptic, (2.5–)3 × as long as broad, narrowed or acuminate at the apex, asymmetric at the base, 4–15 cm. long, 1.3–4 cm. wide, glabrescent apart from the midrib. Male thyrse 4–12 cm. long; female thyrse 1–4 cm. long. Flowers yellow-green or cream; sepals 3–4.

UGANDA. Karamoja District: Moroto Mt., Nakiloro valley, Nov. 1963, *J. Wilson* 1599! & Moroto R., May 1948, *Eggeling* 5789!; Busoga District: Buwenda [Bwenda], Aug. 1940, *Dale* U162!

KENYA. Elgeyo District: Kerio valley, Sept. 1960, *Dale* K1093!; Meru District: Kindani Camp, 20 May 1972, *Ament & Magogo* 305!; S. Kavirondo District: Gucha [Kuja] R., 20 May 1946, *Glasgow* 46/21!

TANZANIA. Biharamulo District: small island off Nyamirembe, Oct. 1957, *Procter* 745!; Handeni District: Kwaluwala, 17 Aug. 1961, *Mgaza* 432!; Kilosa District: Vigude, Kidodi, Nov. 1952, *Semsei* 1021!; Zanzibar, near Chukwani, 26 Nov. 1959, *Faulkner* 2415!

DISTR. **U** 1, 3; **K** 3–5, 7; **T** 1–3, 6; **Z**; Burundi and W. Ethiopia

HAB. Riverine, streamside or lakeside fringing forest, swamp forest, bushland, grassland with scattered trees, sometimes just above high tide mark; also persisting in cultivations; 0–1400 m.

SYN. *L. vaughanii* Dunkley in K.B. 1937: 469 (1937), as '*vaughaniae*'; T.T.C.L.: 558 (1949); I.T.U., ed. 2: 380 (1952)

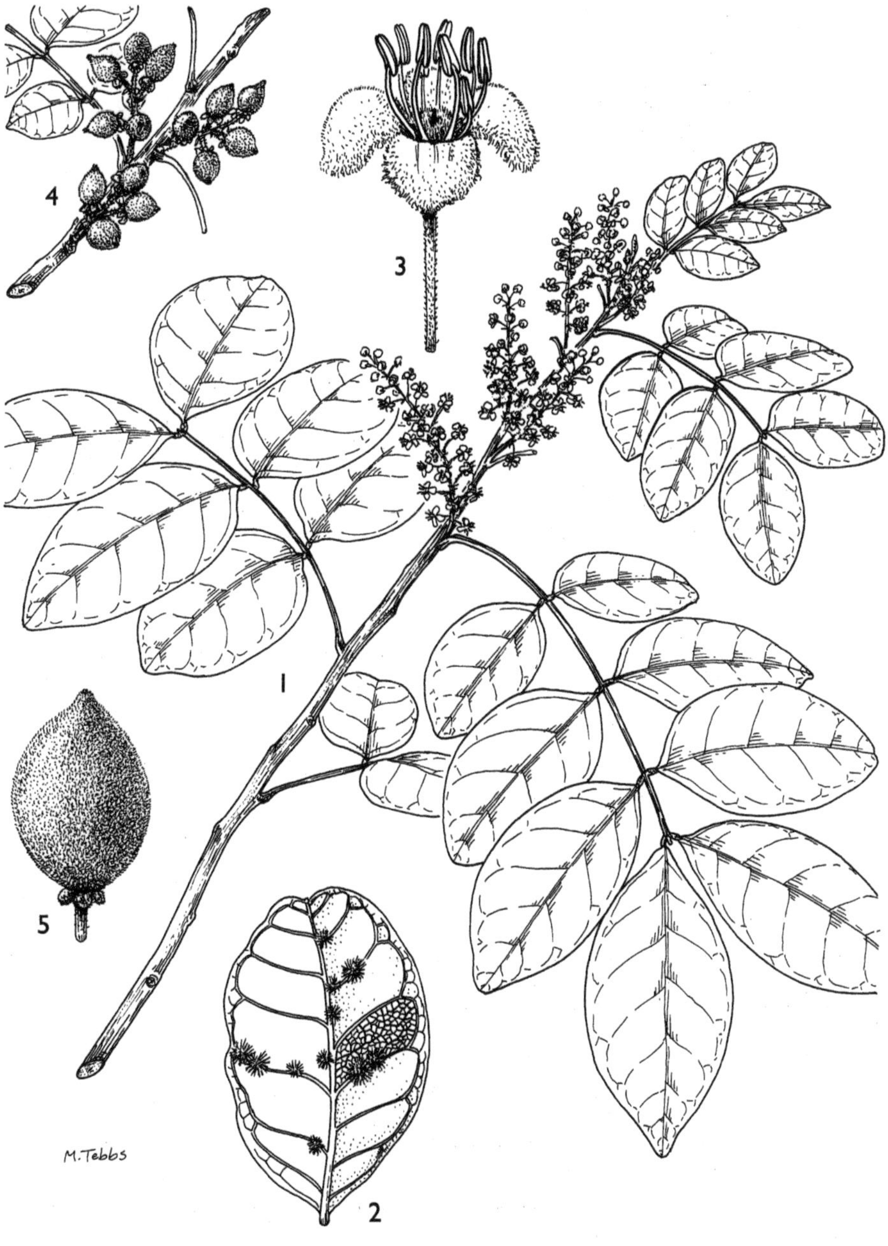

FIG. 13. *LECANIODISCUS FRAXINIFOLIUS* subsp. *SCASSELLATII* — **1**, flowering branch, $\times \frac{2}{3}$; **2**, leaf underside with galls, × 1; **3**, male flower, × 8; **4**, fruiting branchlet, $\times \frac{2}{3}$; **5**, fruit, × 2. 1, 3, 4, from *Polhill* 570; 2, from *Greenway & Kanuri* 13055; 5, from *Faden* 74/987. Drawn by Margaret Tebbs.

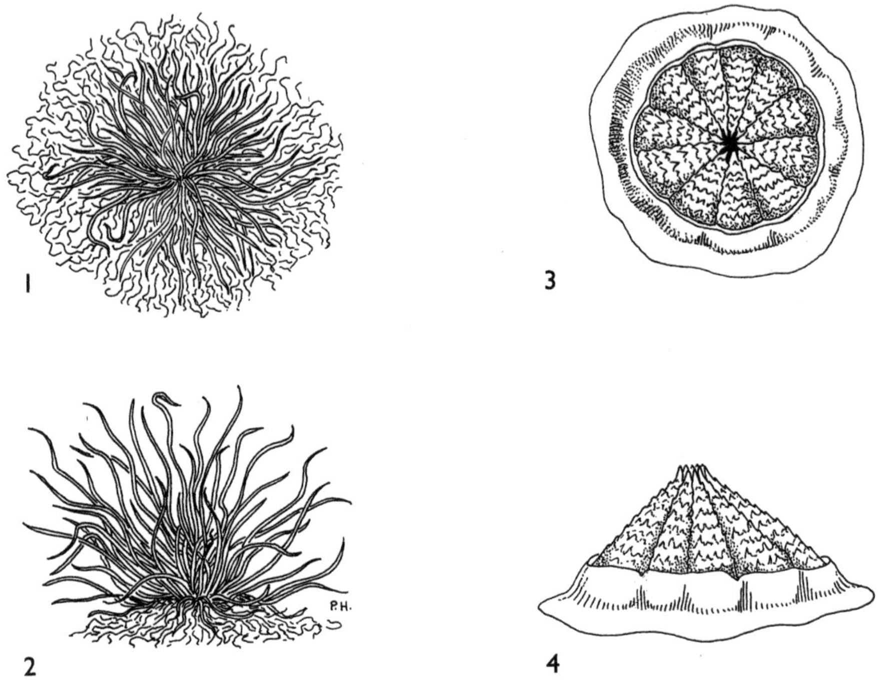

FIG. 14. Galls on *Lecaniodiscus*, all × 12. 1, 2, from *Eggeling* 5789; 3, 4, from *Greenway & Kanuri* 13055. Drawn by Pat Halliday.

subsp. **scassellatii** (*Chiov.*) *Friis* in K.B. 39: 781, map. 1 (1984). Lectotype, chosen by Friis: S. Somalia, Arrar, *Scassellati & Mazzocchi* 2 (FT, lecto., K!, photo.)

Shrub or small tree 2.4–7.5 m. tall; bark brown or silvery grey, smooth or fairly rough. Leaflets pale green, obovate or elliptic, ± twice as long as wide, obtuse or rounded at the apex, cuneate to truncate at the base, 2.3–7 cm. long, 1.5–4.5 cm. wide, glabrous or with scattered glands. Thyrses 0.5–4(–7) cm. long. Flowers white, greenish or greenish yellow, scented; sepals 4. Fig. 13.

KENYA. Teita District: Galana R., Sala Gate campsite, 17 Jan. 1967, *Greenway & Kanuri* 13055!; Kilifi District: Malindi, 23 Dec. 1954, *Verdcourt* 1177!; Lamu District: S. bank of Tana R., Ozi [Osi], 8 Nov. 1957, *Greenway & Rawlins* 9479!

DISTR. **K** 7; S. Somalia

HAB. Coastal forest, bushland, often marginal to swamp forest, riverine forest; 0–150 m.

SYN. *Haplocoelum scassellatii* Chiov., Result. Sci. Miss. Stef.-Paoli, Coll. Bot.: 206 (1916); Radlk. in E.P. IV, 165: 885 (1932)

Chiarinia jubae-fluvii Chiov., Fl. Somala 2: 149, fig. 96, 97 (1932). Type: S. Somalia, Isola, Tonata, *Tozzi* 339 (FT, holo., K!, iso.)

DISTR. (of species as a whole) **U** 1, 3; **K** 3–5, 7; **T** 1–3, 6; **Z**; Burundi, W. Ethiopia, Somalia, Zambia, Malawi, Mozambique and Zimbabwe

NOTE. Subsp. *vaughanii* extends from the coastal forests of S. Kenya and Tanzania inland along the drainage system to around Lake Victoria and W. Ethiopia. It is replaced in the coastal regions of S. Somalia and Kenya by subsp. *scassellatii*, with broader, shorter leaflets. There is overlap on the Kenya coast and inland and possibly the whole population is more or less intermediate and this has led Beentje (K.T.S.L.: 420 (1994)) to dismiss the subspecies as superfluous; in this area there is good reason for such a view but when the whole distribution of the species is examined it is clear there is good reason to recognise subspecies. Examples of intermediates are *Dale* in *F.D.* 3542 (Kwale District, near Mombasa, Mwachi), *Willan* 339 (Kwale District, Diani Beach), *Perdue & Kibuwa* 10068 (Kilifi District, 14 km. S. of Jilore Forest Station) and *Gatheri et al.* 79/59 (Kitui District, Endau). The problem is that the type of *vaughanii* is 'not central' to the narrow-leafleted taxon. There is, however, a vast difference between a tall tree with narrow acuminate leaflets in Karamoja and a small shrubby tree with broad rounded leaflets in N. Kenya and Somalia. Subsp. *fraxinifolius* with leaflets 2–3 times as long as broad, bright green when dry and pilose and glandular beneath occurs in Mozambique, Malawi, Zambia and Zimbabwe.

NOTE ON GALLS. This species is well known for its galls. I am not aware if anyone knows what causes them. There are three types (see fig. 14); one with ± 11 linear-triangular organs protruding from an annulus (according to my colleague B. Spooner caused by a scale insect); another begins as a ball, splits into four parts and a dense tuft of fine rusty filaments is produced (Spooner suggests this is produced by a cecidomyiid midge). The third is similar but smaller. In Uganda and western Kenya and Tanzania only the first occurs but on the coast both this and the second type occur often on the same leaflet but the fluffy one is much more in evidence and sometimes extremely abundant. The third sort is known only in subsp. *fraxinifolius.* The distribution of the galls in the eastern parts of Kenya and Tanzania is inextricably mixed up with the merging of the subspecies and a precise study might throw light on the genetic make up of the populations.

14. **HAPLOCOELUM***

Radlk. in Sitz. Akad. Wiss. München 8: 336 (1878) & in E.P. IV, 165: 881 (1932); Bullock in K.B. 1931: 353–357 (1931)

Pistaciopsis Engl. in E.J. 32: 125 (1902)

Dioecious trees or shrubs with some long shoots and many condensed short shoots. Leaves pinnate; leaflets entire, asymmetric at least at the base, ± notched at apex. Flowers regular in axillary fascicles of cymes, often appearing before the leaves; sepals 5, free; petals lacking; disk hemispherical with the filaments inserted into pockets on its surface. Male flowers: stamens 4–6, much exserted; ovary rudiment absent. Female flowers: staminodes present; ovary ± 3-lobed, with 3-fid style. Fruit coloured, capsule-like, 3-locular, the locules 1-seeded, but mostly indehiscent or irregularly rupturing. Seeds 1–2, covered by a fleshy aril.

A genus of 4 species in Central and eastern tropical Africa extending from Cameroon to Angola, Mozambique and South Africa (Kwazulu-Natal); although formerly 6 species were recorded from East Africa only two are recognised here one having several subspecies. The generic description above does not apply to the endemic Madagascan species *H. perrieri* Capuron which has petals, a bifid stigma, an ovary rudiment in the male flower and densely adpressed fibrous silky fruits. Capuron (Mém. Mus. Nat. Hist. Nat., Paris, 19: 133, t. 26/8–19 (1969)) has erected sect. *Cardiophyllariopsis* for this species. His original intention was to consider it a new genus and I think that might be the best course.

Leaflets in (2–)3–16 pairs, smaller, 1.5–6 × 0.3–0.9 cm.; ovary and fruit round to ovoid or ellipsoid, not angled or obscurely so** . . . . 1. *H. foliolosum*

Leaflets in 1–2(–3) pairs, larger, 0.5–11.5 × 1–5.7 cm.; ovary and young fruit sharply trigonous but mature fruit mostly losing angles or with obscure ribbing . . . . 2. *H. inoploeum*

* By B. Verdcourt. I have diverged so much from the treatment by Bullock which was followed by Davies that it is necessary to claim responsibility.

**Very few female flowers have been collected of *H. foliolosum* and this needs checking so far as the ovary is concerned; one Ugandan specimen has ± angled young fruits.

1. **H. foliolosum** (*Hiern*) *Bullock* in K.B. 1931: 356 (1931); Radlk. in E.P. IV, 165: 1498 (1934); T.T.C.L.: 557 (1949); F.P.S. 2: 341 (1952); C.F.A. 2: 88 (1954); Hauman in F.C.B. 9: 374 (1960); F.F.N.R.: 225 (1962); Exell in F.Z. 2: 530, t. 112 (1966) pro parte; Blundell, Wild Fl. E. Afr.: 137, t. 346 (1987). Type: Angola, Huilla, Lopollo R. and Monino, *Welwitsch* 4507 (BM!, holo., LISU, iso.)

Shrub or tree 4–13.5(–20) m. tall; bark grey, smooth; wood reddish; branchlets dark grey-brown to black, together with the leaf-rhachises grey to golden brown pubescent. Leaves often pinkish when young, 2–10 cm. long, mostly borne on short spur-shoots; petiole 0.3–1.5 cm. long, pubescent; petiolules 0–0.5 mm. long; leaflets in 2–16 pairs, obliquely oblong or elliptic, 1.5–6 cm. long, 3–9 mm. wide (terminal ones largest), obtuse and usually emarginate at the apex, ± truncate or asymmetrically obtuse or cuneate at the base, attached almost at the corner, entire, glabrous to sparsely pubescent mainly on the midrib; lateral nerves numerous, ± closely spaced, venation reticulate. Flowers white, cream or yellow, often appearing before the leaves in axillary clusters or crowded on the spur-shoots; pedicels 1–3 mm. long in male, 5–10 mm. long in female, articulated at the base, pubescent. Sepals 5, oblong, 2 mm. long, 1 mm. wide, pubescent. Stamens 4–5 with filaments 2–5 mm. long. Ovary rounded or somewhat trigonous; style up to 1 mm. long. Fruit orange to red, also described as chestnut or purple, ovoid or subglobose, 1.5–1.8 cm. long, 1.2 cm. wide, apiculate, glabrescent or glabrous. Seeds chestnut, ellipsoid, 1–1.7 cm. long, 6 mm. wide; aril striate.

1. Leaflets mostly in (6–)9–11(–16) pairs, narrowly oblong and square-sided (S. Tanzania) ······· a. subsp. *foliolosum*
   Leaflets mostly in (2–)3–7(–8) pairs with more rounded sides ······· 2
2. Leaflets in 4–7(–8) pairs; style usually ± lacking (Uganda, Kenya and N. Tanzania; 750–1700 m.) ·· b. subsp. *strongylocarpum*
   Leaflets in (2–)3–4 pairs; style persistent, 1–1.5 mm. long (coastal; 0–300 m.) ······· c. subsp. *mombasense*

a. subsp. **foliolosum**

Shrub or small tree 4–6 m. tall. Leaflets mostly in 9–11(–16) pairs, often small and ± square-sided. Style usually very short in fruit.

TANZANIA. Ufipa District: Sumbawanga, Kasanga, near Molwe, 13 Dec. 1948, *Glover* in *Bredo* 6422!; Dodoma District: Kazikazi, Ulimiri Thicket, 11 July 1932, *B.D. Burtt* 3795! & Itigi–Singida road, Itigi Thicket, 26 Mar. 1965, *Richards* 19931! & Namanzi, 27 Oct. 1973, *Ruffo* 1475!

DISTR. **T** 4, 5, (6, 7, see note); Zaire, Angola, Zambia, Mozambique and Zimbabwe

HAB. Deciduous *Pseudoprosopis* thickets, woodland; 1200–1350 m.

SYN. ?*Balsamea foliolosa* Hiern, Cat. Afr. Pl. Welw. 1: 126 (1896)
Pistaciopsis dekindtiana Engl. in E.J. 32: 126 (1902). Types: Angola, Huila, Mucha Mts., *Dekindt* 216 (B†, holo.)
Haplocoelum dekindtianum (Engl.) Radlk. in E. & P. Pf., Nachtr. 3: 204 (1907) & in E.P. IV, 165: 885 (1932)

NOTE. The type has 6–10 pairs of leaflets, more rounded at the sides. Angolan material can have ± distinct styles. Burtt (3556) preserved seedlings with 2–4 pairs of leaflets which he had germinated from civet cat dung at Kazikazi. Two specimens have been seen from vegetation described as 'submontane rain-forest' and '*Parinari/Piptadenia* forest' at 1200–1400 m., Morogoro District, Uluguru Mts., NNE. valley of Lupanga, 28 Sept. 1988, *Pócs & Knox* 88192/0; Iringa District, Vigola, 2 Oct. 1958, *Ede* 52. They are trees 6–13.5 m. tall. The leaflets are distinctive, the apical notch more pronounced than usual, up to 1 mm. deep and the upper leaflets are more narrowly cuneate at the base. They may represent a further subspecies but the material is sterile.

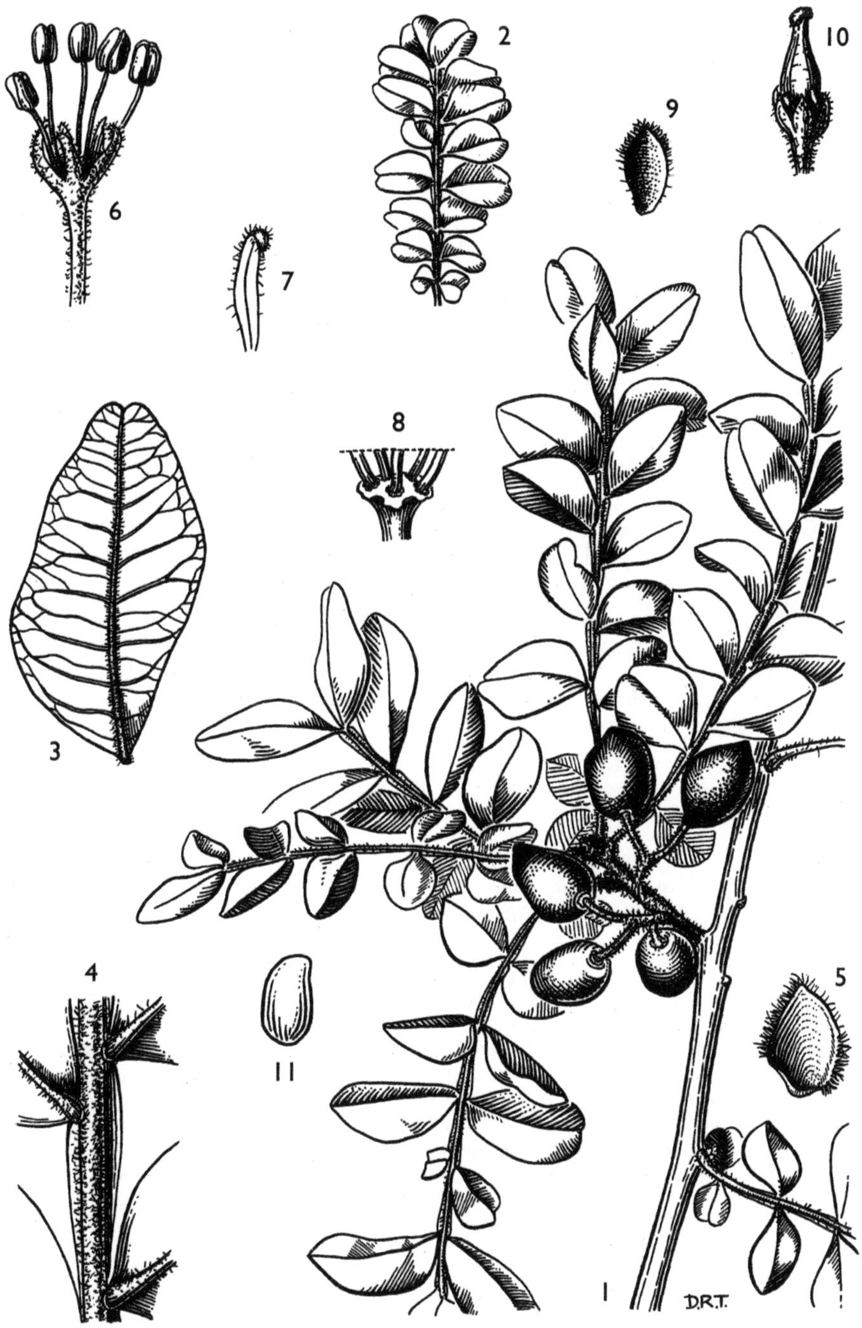

FIG. 15. *HAPLOCOELUM FOLIOLOSUM* subsp. *STRONGYLOCARPUM* — **1**, fruiting branch, × 1; **2**, leaf (to show variation in size), × 1; **3**, leaflet, × 2; **4**, portion of leaf-rachis, × 4, **5**, bract, from base of pedicel, × 8; **6**, male flower, × 4; **7**, sepal from male flower, × 8; **8**, receptacle, showing position of stamens, × 10; **9**, sepal from female flower, × 8; **10**, pistil, × 4; **11**, seed, × 1. 1, from *B.D. Burtt* 6459. Drawn by D.R. Thompson.

subsp. **strongylocarpum** (*Bullock*) *Verdc.* comb. et stat. nov. Type: Kenya, Kisumu, *Dobbs* in *F.D.* 2364 (K!, holo., K!, iso.)

Shrub or tree 2.7–10(–18) m. tall, with smooth grey bark flaking to show pale patches; slash yellow. Leaflets in 4–7(–8) pairs, asymmetrically elliptic-oblong, 0.8–3.2 × 0.7–1.6 cm., emarginate. Style absent or very short. Fig. 15.

UGANDA. W. Nile District: E. Madi, Umi, near R. Nile, Mar. 1935, *Eggeling* 1775!; District uncertain: Lake Albert Rift, 1 Oct. 1935, *Hancock* 102/36!; Busoga District: N. side of Sigulu I., Matolo Landing, 14 Jan. 1953, *G.H.S. Wood* 595!

KENYA. Northern Frontier Province: Moyale, 29 Sept. 1952 & 20 Oct. 1952, *Gillett* 13972!; Machakos District: Nairobi–Mombasa road, near Ulu, 1 May 1970, *Gillett* 19121! & 19122!; Masai District: Namanga, foot of Ol Donyo Orok, 27 Mar. 1952, *Bally* 8102!

TANZANIA. Mwanza District: Uzinza, Geita–Karumo road, Ibanda Hill camp, 8 Apr. 1937, *B.D. Burtt* 6459! & Lake Victoria, Rubondo I., 27 Oct. 1956, *Gane* 83!; Musoma District: Serengeti National Park, Kirawira guard post, 30 Dec. 1931, *Greenway & Myles Turner* 14946!

DISTR. **U** 1–4; **K** 1, 2 (*fide* Beentje), 3–6, 7 (Teita); **T** 1, ?2, ?3; Cameroon, Central African Republic, Rwanda, Ethiopia and ?S. Somalia

HAB. Grassland with *Combretum* etc., dry thicket and bushland, particularly on rock outcrops, rocky mountains with quartzite sand and pebbles, riverine forest, also often a dominant on lake shore soils; 750–1700 m.

SYN. *Pistaciopsis gallaensis* Engl. in E.J. 32: 125 (1902). Type: Ethiopia, Galla Highlands, near Ginea-Dumek, *Ellenbeck* 1963 (B†, holo.)

Haplocoelum gallaense (Engl.) Chiov., Result. Sci. Miss. Stef.-Paoli, Coll. Bot.: 207 (1916); Radlk. in V.E. 3(2): 278 (1921); Bullock in K.B. 1931: 356 (1931); Radlk. in E.P. IV, 165: 885 (1932); T.T.C.L. 558 (1949); Fouilloy & Hallé in Fl. Cameroun 16: 145 (1973) & in Fl. Gabon 23: 145 (1973); Troupin, Fl. Rwanda 2: 304, fig. 95/2 (1983)

H. strongylocarpum Bullock in K.B. 1931: 356 (1931); T.T.C.L.: 558 (1949)

[*H. foliolosum* sensu I.T.U., ed. 2: 379 (1952); K.T.S.: 512, fig. 93 (1961); Vollesen in Fl. Ethiopia 3: 507 excl. fig. (1990); Beentje, K.T.S.L.: 419, fig. (1994), *non* (Hiern) Bullock sensu stricto]

subsp. **mombasense** (*Bullock*) *Verdc.* in K.B. 51: 802 (1996). Type: Kenya, Mombasa, Nov. 1884, *Wakefield* (K!, holo.)

Small tree or twiggy shrub 2–10 m. tall with grey finely grooved bark. Leaflets in (2–)3–4 pairs, obovate-oblanceolate, elliptic, obovate or almost round, 0.6–3.2 cm. long, 0.35–1.8 cm. wide, rounded and emarginate at the apex. Ovary globose. Fruit globose, 0.8–1 cm. long; style persistent, 1–1.5 mm. long.

KENYA. Kwale District: Mrima Hill to Marenje Forest, 8 Mar. 1977, *R.B. & A.J. Faden* 77/722!; Kilifi District: Kaya Dagamra, 23 Apr. 1989, *S.A. Robertson & Luke* 5719!; Tana River District: TARDA compound, Gamba, 21 Jan. 1991, *S.A. Robertson* 6435!

TANZANIA. Handeni District: Mgera, 15 July 1960, *Semsei* 3037!; Tanga District: N. of Ngole, 9 June 1937, *Greenway* 4949! & 4950!; Uzaramo District: Pugu Hills, NW. of Kiserawe, 9 July 1982, *Hawthorne* 1098! & ± same locality, 24 Oct. 1968, *B.J. Harris* 2475!

DISTR. **K** 7; **T** 3, 6, 8; S. Somalia, Mozambique & probably South Africa (Kwazulu-Natal)

HAB. Hillside grassland, *Acacia-Combretum* woodland, relict floodplain forest, seashore bushland, *Avicennia* swamp edges, also lowland *Brachystegia-Isoberlinia* woodland; 0–300 m.

SYN. *H. mombasense* Bullock in K.B. 1931: 355 (1931); T.T.C.L.: 558 (1949); K.T.S.: 512 (1961); Vollesen in Opera Bot. 59: 58 (1980); Beentje, K.T.S.L.: 419 (1994)

NOTE. This species would be an ideal subject for ecological investigation of the variations.

2. **H. inoploeum** Radlk. in Sitz. Akad. Wiss. München 8: 337 (1878) & in P.O.A.C.: 251 (1895) & in V.E. 3(2): 278 (1921); Bullock in K.B. 1931: 355 (1931); Radlk. in E.P. IV, 165: 883 (1934); T.T.C.L.: 558 (1949); K.T.S.: 512 (1960); Vollesen in Opera Bot. 59: 58 (1980); Beentje, K.T.S.L.: 419, map (1994). Type: Kenya, Mombasa, 1847–1852, *Boivin* (P, holo.)

Semideciduous or ± evergreen shrub or small tree 2.4–15 m. tall, occasionally reported to be subscandent with stout interlacing grey-brown branches, with shoots sparsely to densely brownish-pubescent at first, later glabrescent; bark grey, grey-brown or blackish, smooth to rough, somewhat fissured, flaking or peeling in irregular patches; wood very hard. Leaves tufted on short shoots; rhachis 0.4–3(–4) cm. long, narrowly winged, sparsely to densely pubescent; petioles 1–1.1 cm. long; petiolules 0.1–2 mm. long, pubescent; leaflets yellow-green in 1–2(–3) pairs, opposite or not, oblong, ovate, round or elliptic, 0.5–11.5 cm. long, 1–5.7 cm. wide, obtuse and ± emarginate at the apex, asymmetrically obtuse or acute at base with one side convex and the other concave, glabrous; lateral nerves in 8–15 pairs, the venation finely reticulate above. Flowers white, cream, yellow or greenish, sweet-scented in clusters of cymes in the leaf-axils, or often appearing on short shoots before the leaves; pedicels 2–9 mm. long. Male flowers: sepals linear-lanceolate or oblong, 1–1.5 mm. long, pubescent; stamens 5, much exserted, glabrous. Female flowers: staminodes short; ovary very acutely trigonous, glabrous; stigma 3-fid. Fruit red, orange or yellow, ellipsoidal, at first trigonous, usually becoming rounded at maturity but some retain 2–3 blunt keels particularly near the style base, 1.2–2 cm. long, 0.6–1.2 cm. wide, acuminate, glabrous, not truly dehiscent but breaking irregularly. Seeds brown, irregularly ovoid or ellipsoid, 7.5–13 mm. long, 4–9 mm. wide.

KENYA. Kwale District: Jadini Hotel grounds, 3 Dec. 1959, *Greenway* 9622!; Kilifi District: Arabuko-Sokoke Forest, 21 Apr. 1985, *S.A. Robertson* 3866!; Tana River District: 48 km. S. of Garsen, Kurawa, 28 Oct. 1961, *Polhill & Paulo* 671!

TANZANIA. Tanga District: Sawa, 3 Apr. 1956, *Faulkner* 1831! & 1832!; Pangani District: 30 km. SW. of Pangani, Msubugwe Forest Reserve, 17 Mar. 1950, *Verdcourt* 118!; Uzaramo District: 11 km. N. of Dar es Salaam, Kawe, 15 Jan. 1971, *B.J. Harris & Jones* 5550!; Pemba, Pembe I., 11 Oct. 1929, *Vaughan* 721!

DISTR. **K** 7; **T** 3, 6, 8; **P**; S. Somalia

HAB. Coastal forest edges, semi-evergreen bushland and thicket on coral rag, dunes and rock outcrops; also wooded grassland, *Brachystegia* woodland, *Brachylaena* forest and swamp and riverine fringes; 0–600 m.

SYN. *Pistaciopsis wakefieldii* Engl. in E.J. 32: 125 (1902) & in E.J. 34: 157 (1904). Type: Kenya, Mombasa, *Wakefield* pro parte (B†, holo., K!, iso.)

Haplocoelum trigonocarpum Radlk. in Sitz. Akad. Wiss. München 20: 249 (1890) & in P.O.A. C: 251 (1895); Bullock in K.B. 1931: 355 (1931); Radlk. in E.P. IV, 165: 884 (1932); T.T.C.L.: 558 (1949); K.T.S.: 514 (1961); Vollesen in Opera Bot. 59: 58 (1980); Beentje, K.T.S.L.: 419 (1994). Type: Tanzania, Uzaramo District, Dar es Salaam*, *Kirk* (K!, lecto.)

H. wakefieldii (Engl.) Chiov., Result. Sci. Miss. Stef.-Paoli, Coll. Bot.: 207 (1916)

Filicium somalense Chiov., Fl. Somala 2: 143, fig. 94 (1932). Type: Somalia, Bur Gao, *Senni* 110 (FT, holo.)

NOTE. *H. inoploeum* and *H. trigonocarpum* have always been maintained as separate species one having ovoid fruits, the other trigonous ones. The identical facies of the two supposed species and the absence of any female flowering specimens with ovoid ovaries (which might have been expected in an ovoid-fruited species) aroused my suspicions. *Faulkner* 1832 has both ovoid and trigonous fruits mixed and it is clear that the angles are lost when the fruit is completely ripe or only obscure ribs remain. Many specimens have the ribbing obscure and were treated as dubiously one species or the other and all specimens with female flowers were automatically determined as *H. trigonocarpum* since the ovary is always trigonous. Male specimens were sorted about equally between the two. The hard wood is used for walking sticks and clubs.

* The persistent error of botanists to interpret Zanguebar as only Zanzibar I. has resulted in this type being attributed to the latter. During most of the 19th century Zanguebar extended for at least ten miles inland from the mainland coast.

15. **CAMPTOLEPIS**

Radlk. in E. & P. Pf., Nachtr. 3: 207 (1907) & in E.P. IV, 165: 1326 (1933); Capuron in Mém. Mus. Nat. Hist. Nat., Paris, 19: 108 (1969); Friis & Vollesen in K.B. 40: 860 (1985)

Hypseloderma Radlk. in E.P. IV, 165: 886 (1932)

Small trees. Leaves pinnate; leaflets entire, leathery, glabrous. Inflorescence borne on the previous season's wood, consisting of axillary fascicles of short cymes. Flowers regular; sepals 5, petaloid, concave; petals 5, of unequal size, scales present, entire to completely divided into 2; disk lobed, glabrous. Stamens 5–13. Ovary 3–5-locular. Fruit (where known) a drupe or berry, indehiscent. Seeds 1–3, enclosed in a fleshy aril.

A genus of 4 species in Madagascar and one also extending to the E. African coast.

C. ramiflora (*Taub.*) *Radlk.* in E. & P. Pf., Nachtr. 3: 207 (1907) & in E.P. IV, 165: 1327 (1933); T.T.C.L.: 556 (1949); Verdc. in K.B. 13: 218 (1958); Capuron in Mém. Mus. Nat. Hist. Nat., Paris, 19: 112, t. 27/5–10 (1969); Friis & Vollesen in K.B. 40: 860 (1985); Beentje, K.T.S.L.: 415 (1994). Type: Tanzania, Bagamoyo District, *Stuhlmann* 142 (B†, holo.); Bagamoyo, Feb. 1889, *Sacleux* 682 (P!, neo., P!, isoneo., K!, photo.)

Dioecious, much-branched small tree, ± 3 m. tall in Tanzania but 12–30 m. tall in Somalia and Madagascar; young twigs densely pubescent at first, later glabrous, with bark whitish, smooth, later fissuring vertically to become red-brown; older bark peeling like a plane tree. Leaf-rhachis 2–8(–12.5) cm. long, somewhat flattened and with raised lateral lines, pilose or glabrous; petiolules 2–8 mm. long, channelled, swollen; leaflets in 2–4(–5) pairs, opposite, ovate to narrowly elliptic, 0.5–1.4 cm. long, 1.5–4 cm. wide, the uppermost pair largest, the lowermost slightly to $^1/_3$ smaller, shortly and bluntly acuminate, the base cuneate and usually asymmetric, thick, leathery, drying purplish-brown beneath, glabrous apart from scattered hairs on the midrib; midrib sunken above, prominent beneath, lateral nerves looping inside the margin with smaller veins between the loops and the margins, venation raised and reticulate beneath. Flowers 5–6 mm. long, borne in fascicles of cymes; inflorescence-axes 1.5–6 cm. long, the branches ± 1 cm. long, both yellowish pubescent, abruptly transitional to glabrous; pedicels 1–2 mm. long, deciduous with the (male) flowers. Sepals unequal, orbicular, 2–4 mm. long, concave, glabrous; petals 5, white, slightly exceeding the sepals, ovoid, briefly clawed, 4 mm. long, 3.5 mm. broad, the scale half as long as the limb, villous-hairy. Stamens 12, shortly exserted; anthers oblong, 1.3 mm. long, briefly auriculate at the base; filaments flattened, glabrous. Ovary rudiment minute. Female flowers and fruits not seen (see note). Fig. 16.

KENYA. Lamu District: Witu Forest, 15 Nov. 1988, *S.A. Robertson & Luke* 5476! & Witu Forest Reserve, cultivated in Mombasa, Bamburi Nature Park model Kaya, 6 Mar. 1997, *S.A. Roberston & Baer* 7229!

TANZANIA. Bagamoyo, Feb. 1889, *Sacleux* 682!; Uzaramo District: Msasani, 2 Jan. 1974, *Ruffo* 1009! & 18 Feb. 1939, *Vaughan* 2757!

DISTR. **K** 7; **T** 6; Somalia and Madagascar

HAB. *Trichilia, Garcinia, Hyphaene* forest (in Kenya); 10–100 m.

SYN. *Deinbollia ramiflora* Taub. in P.O.A. C: 250 (1895)

Haplocoelum jubense Chiov., Result. Sci. Miss. Stef.-Paoli, Coll. Bot.: 56, 206 (1916); Bullock in K.B. 1931: 357 (1931). Types: Somalia, Juba R., Duchia, *Paoli* 464 & 465 (FT, syn., K!, photo.) & Revai, *Paoli* 477 (FT, syn., K!, photo.) & Uasciani, *Scasselati & Mazzocchi* 76 (FT, syn., K!, photo.)

Hypseloderma jubense (Chiov.) Radlk. in E.P. IV, 165: 886 (1932); Friis in K.B. 39: 783, map 1 (1984)

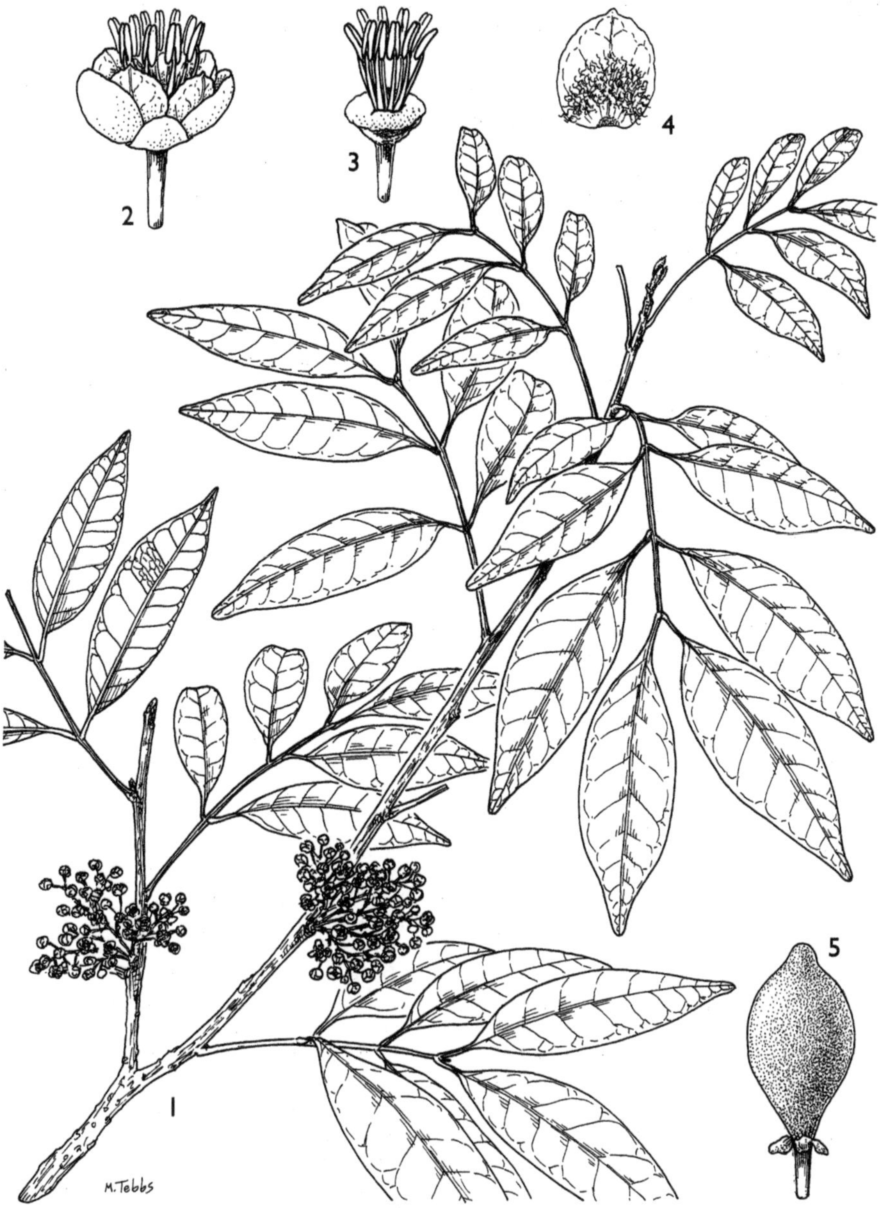

FIG. 16. *CAMPTOLEPIS RAMIFLORA* – **1**, flowering branch, × $^{2}/_{3}$; **2**, male flower, × 4; **3**, male flower with sepals and petals removed, × 4; **4**, petal, × 4; **5**, fruit, × 2. 1–4, from *Ruffo* 1009; 5, from photograph of *Paoli* 477. Drawn by Margaret Tebbs.

NOTE. Chiovenda described the ovary as globose, 3–4 mm. long and wide with stipe 1.2 mm. long; stamens (staminodes) ?10 with filaments 2 mm. long, glabrous; ovary 3-locular (*fide* Capuron); fruit black, indehiscent, spherical with an apical areole around the style remnant, 1.2–2.5(–3) cm. long. Seeds (*fide* Chiovenda) obovate, 1.5 cm. long, 9 mm. wide, 7 mm. thick, completely surrounded by a fleshy aril; Capuron states hilum placed at base of seed on adaxial side, ± vertical.

16. **LEPISANTHES**

Blume, Bijdr.: 237 (1825); Radlk. in E.P. IV, 165: 726 (1932); Leenh. in Blumea 17: 33–91 (1969) & in Fl. Males., ser. 1, 11: 627 (1994)

Aphania Blume, Bijdr.: 236 (1825); Radlk. in E.P. IV, 165: 699 (1932)

Trees or shrubs. Leaves pinnate; leaflets opposite or alternate, the lowermost often reduced and sometimes stipule-like. Inflorescences terminal and axillary, paniculate or simple. Flowers regular; sepals 4–5, petaloid, unequal, free, strongly imbricate; petals (0–)4–5, scale simple or complicated; disk complete, convex. Stamens 4–18, mostly ± 8. Ovary 2–3-lobed with 1 ovule per lobe; style entire. Fruit of 1–3 mericarps, indehiscent. Aril absent.

A genus redefined by Leenhouts to comprise 24 species; mainly in Malesia but one also extending to India and Africa. Several genera previously kept separate are reduced to sections of *Lepisanthes*.

L. senegalensis (*Poir.*) *Leenh.* in Blumea 17: 85 (1969); Vollesen in Opera Bot. 59: 58 (1980); Blundell, Wild Fl. E. Afr.: 137, t. 217 (1987); Vollesen in Fl. Ethiopia 3: 504, fig. 125.10 (1990); Beentje, K.T.S.L.: 420, fig (1994); Leenh. in Fl. Males., ser. 1, 11: 651, fig. 52a (1994). Type: Senegal, *Adanson & Geoffroi f.* (P, holo., K!, ?iso.)

Monoecious evergreen tree, 6–21 m. tall with heavy crown; bark grey or pale brown, smooth; slash pale orange. Foliage with petioles etc. and base of midribs crimson when young; leaf-rhachis 0–10 cm. long, including 0–5 cm. long petiole; petiolules 0–1 cm. long; leaflets in 1–3 pairs (sometimes reduced to 1 leaflet), the lowest pair not noticeably reduced, the upper pair opposite, drying greyish green above and paler green beneath, elliptic, 6.5–27 cm. long, 2–10 cm. wide, rounded to acuminate at the apex, cuneate at the base, entire, glabrous; secondary nerves in 12–20 pairs, ± indistinct. Inflorescence said to be ± unpleasantly scented, often appearing terminal and accompanying a new flush of leaves, paniculate, 5–20 cm. long, 5–15 cm. wide, the branches thyrsoid with both clusters and single flowers; axes at first golden pubescent, elongating in fruit; pedicels 0–3 mm. long, enlarging to 7 mm. in fruit. Flower greenish yellow; sepals 5, the outer 2 shorter than the inner, ovate or round, 2–3.5 mm. long, 1.5–3 mm. wide, glabrous apart from the white margin; petals 5 (rarely absent), ovate, 2–3 mm. long and wide, clawed, scale half as long as the limb, fringed with hairs; disk pentagonal, 3.6 mm. in diameter, lobed. Stamens 5–7, hairy at base, glabrous for most of length. Ovary 2(–3)-lobed; style entire; ovary rudiment in male flowers a cushion of hairs. Mericarps crimson, 2(–3), developed ones ellipsoid, 1–1.8 cm. long, 0.8–1.5 cm. wide, but 1–2 abortive at base. Seeds black, ± 1.2 cm. long, 9 mm. wide. Fig. 17.

UGANDA. W. Nile District: Zoka Forest, Jan. 1952, *Leggat* 45!; Acholi District: Chua, Lututuru, Feb. 1938, *Eggeling* 3513!; Ankole District: Lake Lutoto, 4 Jan. 1953, *Osmaston* 2754!
KENYA. W. Suk District: Sebit, Dec. 1964, *Tweedie* 2961!; Masai District: Mara Masai Reserve, Talek [Telek] R., 12 Sept. 1947, *Bally* 5331!; Lamu District: Utwani, 1 Mar. 1957, *Rawlins* 353!
TANZANIA. Musoma District: source of Bolgonja [Bolongonya] R., 13 Nov. 1953, *Tanner* 1833!; Moshi District: 1.5 km. W. of Moshi, 25 Feb. 1953, *Drummond & Hemsley* 1320!; Rufiji District: Mafia I., Bweni–Mawalani, 17 Aug. 1937, *Greenway* 5134!; Kilwa District: Selous Game Reserve, Lungonya Forest, 26 Sept. 1970, *Ludanga* 1140!

FIG. 17. *LEPISANTHES SENEGALENSIS* — **1**, flowering branchlet, × 1; **2**, leaf, much reduced; **3**, bract, × 8; **4**, male flower, × 4; **5**, **6**, sepal, × 4; **7**, petal, × 4; **8**, male flower with sepals and petals removed, × 4; **9**, ovary rudiment in male flower, × 8; **10**, female flower with sepals and petals removed, × 4; **11**, fruit, × 1. Drawn by D.R. Thompson.

DISTR. *U 1–4; **K** 1–7; **T** 1–3, 6, 8; west tropical Africa, Ethiopia, S. Somalia and Mozambique, also India and Malesia

HAB. Evergreen lowland and submontane forest, evergreen forest, often on coral and lava rock; 0–1800 m.

SYN. *Sapindus senegalensis* Poir. in Lam., Encycl. Méth. Bot. 6: 666 (1804); F.T.A. 1: 430 (1868)
Ornitrophe thyrsoides Schumach. & Thonn., Beskr. Guin. Pl.: 185 (1827). Type: Ghana, Quitta, *Thonning* 342 (C, holo.)
Sapindus abyssinicus Fresen. in Mus. Senckenb. 2: 277 (1837). Type: Ethiopia, Simien Mts., Kulla, 2 days N. of Gondar, *Fresenius* (FR, holo.)
Schmidelia thyrsoides (Schumach. & Thonn.) Baker in F.T.A. 1: 423 (1868)
Aphania senegalensis (Poir.) Radlk. in Sitz. Akad. Wiss. München 8: 307 (1878) & in E.P. IV, 165: 703 (1932); F.P.N.A.: 522 (1948); T.T.C.L.: 555 (1949); I.T.U., ed. 2: 377 (1952); F.P.S. 2: 336 (1952); E.P.A. 1: 492 (1958); F.W.T.A., ed. 2, 1: 716 (1958); Hauman in F.C.B. 9: 343, t. 36 (1960); K.T.S.: 507 (1961); Breitenb., Indig. Trees Ethiopia: 239 (1963); Exell in F.Z. 2: 525, t. 108 (1966); Hamilton, Field Guide Uganda For. Trees: 225, fig. 420 (1981); Fouilloy & Hallé in Fl. Cameroun 16: 75, t. 15/1–9 (1973) & in Fl. Gabon 23: 75, t. 15/1–9 (1973)
A. silvatica Hutch. & Dalziel, F.W.T.A. 1: 502 (1928); E.P. IV, 165: 705 adnot. (1932). Types: Ghana, Tano R., *Chipp* 345 & *Chevalier* 20064 (K!, syn.)

NOTE. In Malesia a large number of morphological races exist with different sizes of fruit, the races being correlated with different habitats (Leenh., 1969 & 1994, with extensive extra-African synonymy). Hauman (1960) claims the fruit is poisonous on account of the saponins it contains and Leenhouts says the fruit is edible but the seeds probably poisonous in quantity. There is a definite record of 14 goats being poisoned near the Tana R. The wood has been used for poles and for making small utensils; the grain is attractive. The unijugate form is rather distinctive and Gillett (adnot. K) has suggested it could be recognised. All material from **K** 1, 2, 4, 7, **T** 2, 6, 8 and Mozambique is unijugate, both occur in **K** 5, and in **K** 6 and **T** 1 all are 2–3-jugate. Here is an eco-geographical problem which could be investigated in the field.

17. GLENNIEA

Hook.f., G.P. 1: 404 (1862); Radlk. in E.P. IV, 165: 858 (1932), as '*Gleniea*'; Leenh. in Blumea 22: 411–414 (1975) & in Fl. Males., ser. 1, 11: 540 (1994)

Crossonephelis Baill. in Adansonia 11: 245 (1874); Radlk. in E.P. IV, 165: 818 (1932); Leenh. in Blumea 21: 91–103 (1973)
Melanodiscus Radlk. in T. Durand., Ind. Gen. Phan.: 75 (1888) & in E.P. IV, 165: 816 (1932)

Trees or shrubs or occasionally scandent, monoecious or dioecious. Leaves simple or pinnate; leaflets in 1–6 pairs. Inflorescences terminal and in upper leaf axils, thyrsoid or paniculate with few spreading branches. Flowers regular; sepals 4–5, ± free; petals 5 or lacking; disk entire, broad, lobed. Stamens 4–8. Ovary 2–3-lobed; ovules 1 per locule; style 2–3-lobed. Fruit of 1–3 mericarps. Seeds without arils.

Eight species in Malesia, Sri Lanka, Madagascar and tropical Africa; one is widespread in central and eastern Africa.

G. africana (*Radlk.*) *Leenh.* in Blumea 22: 412 (1975); Beentje, K.T.S.L.: 418, fig. (1996), as '*Glenniaea*'. Type: Nigeria, Lagos, *Moloney* (K!, holo.)

Dioecious deciduous shrub or tree, (3–)7–24 m. tall; bark grey, smooth; slash orange. Leaves yellow when young, turning pink-flushed then pale green; leaf-rhachis 3–12 cm. long, the lowest leaflets near the base forming pseudo-stipules

* I have seen no material from **U** 3, **K** 3 or **T** 3. [B.V.]

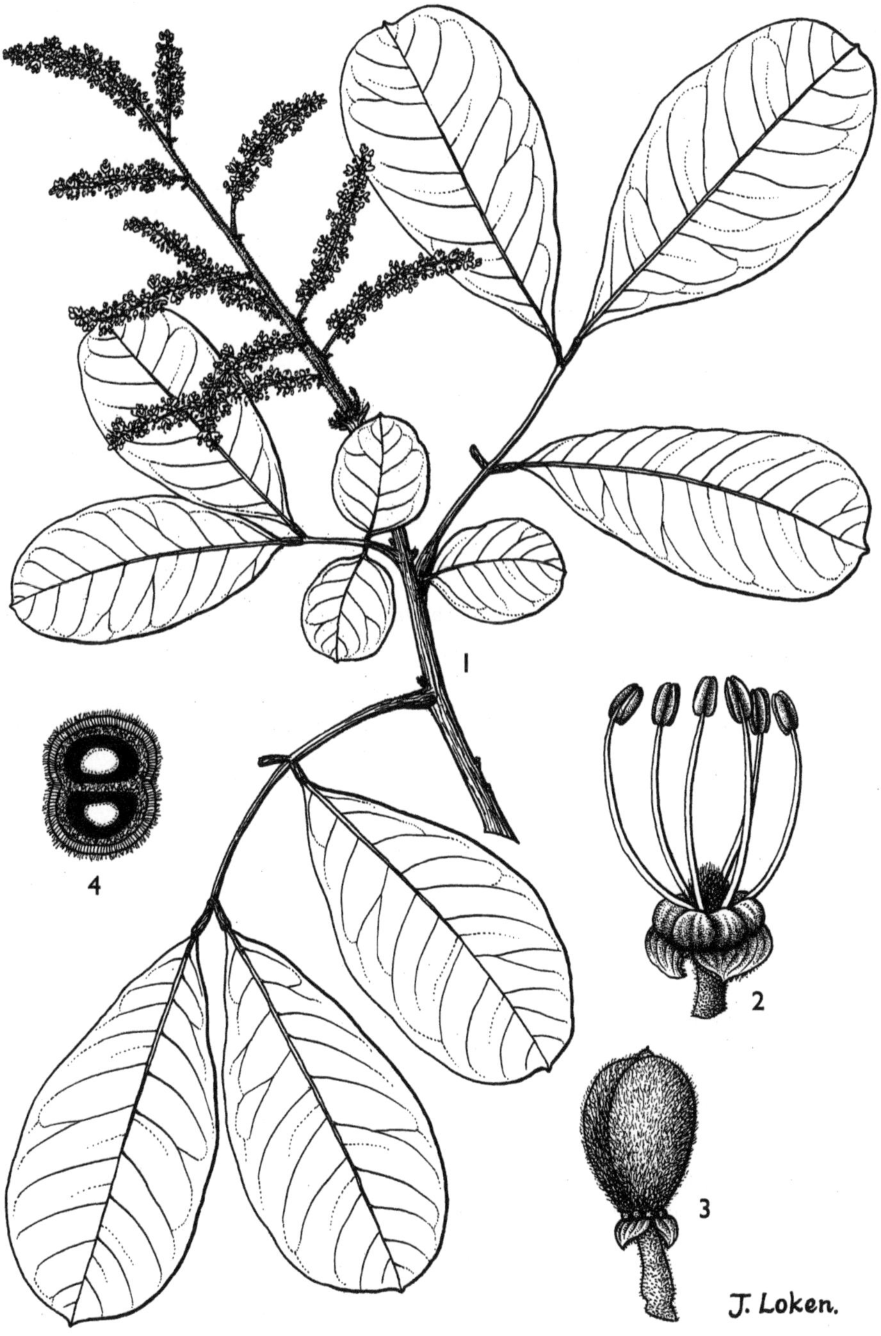

FIG. 18. *GLENNIEA AFRICANA* — **1**, flowering branchlet, × ⅔; **2**, male flower, × 5; **3**, young fruit, × 3; **4**, transverse section of fruit, × 3. 1, from *Chase* 4523; 2, from *Chase* 4419; 3, 4, from *Chase* 4697. Reproduced from F.Z. 2, t. 110 (1966).

1.5–4 cm. long; leaflets in (1?–)2–3(–4?) pairs, oblanceolate, elliptic, ovate or lanceolate, 3–25 cm. long, 1.5–11 cm. wide, the uppermost the largest, the lowermost much smaller, acute or obtuse at the apex, cuneate to truncate at the base, the margin sometimes undulate, glabrous save for main nervation beneath; main lateral nerves in 4–14 pairs. Inflorescences terminal, axillary or cauliflorous; main axis 5–15 cm. long, elongating to 30 cm. in fruit, usually branched, the branches 3–5 cm. long, pseudo-racemose with flowers single or in short cymes; pedicels 2–9 mm. long. Flowers brownish yellow, yellowish white, pinkish yellow or brownish red; sepals 4, ovate, 1.5–2 mm. long, spreading, pubescent outside, nearly glabrous inside; petals lacking; disk yellow. Stamens 5, well exserted; filaments 3–8 mm. long, glabrous; staminodes with filaments 1 mm. long. Ovary yellow-pubescent; style 2(–3)-lobed, glabrous, drying dark coloured. Fruit orange, drying ferruginous pubescent, of 1–2 mericarps, single ones usually asymmetrically ovoid, 2–2.5 cm. long; pulp fleshy. Fig. 18.

UGANDA. W. Nile District: Zoka Forest, Jan. 1952, *Leggat* 77!; Bunyoro District: Budongo Forest, Feb. 1935, *Eggeling* 1509!; Busoga District: Btembe Bunya, 1.5 km. N. of Lubanyi [Lubani] Hill, 8 Nov. 1950, *G.H.S. Wood* 12!

KENYA. Machakos District: Kibwezi, Aug. 1938, *Teofilo* in *C.M.* 7601! & Aug. 1936, *Gardner* in *F.D.* 3710!; 'Coast district', 24 June 1909, *Battiscombe* 57!

TANZANIA. Moshi District: Rau Forest, 25 Sept. 1924, *Lewis* 236!; Tanga District: Mkulumuzi valley, near the Amboni Caves, 8 June 1950, *Verdcourt & Greenway* 258!; Morogoro District: Mindu Mt., 29 June 1983, *Polhill, J.C. & J.M. Lovett* 4947a!; Zanzibar, near Kutuoni, 5 July 1931, *Vaughan & Jonah* 1942!

DISTR. ***U** 1–4; **K** 2, 4, 7; **T** 2, 3, 6; **Z**; Nigeria, Cameroon, Zaire, Sudan, Malawi, Mozambique and Zimbabwe

HAB. Wet evergreen forest, riparian forest and bushland, sometimes on limestone; 30–2000 m.

SYN. *Melanodiscus africanus* Radlk. in T. Durand., Ind. Gen. Phan.: 75 (1888) & in E.P. IV, 165: 817 (1932); T.T.C.L.: 559 (1949); F.W.T.A., ed. 2, 1: 720 (1958)

M. oblongus Radlk. in P.O.A. C: 250 (1895) & in E & P. Pf. III, 5: 32 (1895) & in E.P. IV, 165: 817 (1932); T.T.C.L.: 559 (1949); Hauman in F.C.B. 9: 366 (1960); K.T.S.: 515 (1961); Exell in F.Z. 2: 528, t. 110 (1966). Type: Tanzania, Kilimanjaro, *Kersten* (B†, holo.)

M. sp. nov.? sensu I.T.U.: 217 (1940) & ed. 2: 381 (1952); Hamilton, Field Guide Uganda For. Trees: 225, fig. 418 (1981)

Crossonephelis africanus (Radlk.) Leenh. in Blumea 21: 95 (1973); Fouilloy & Hallé in Fl. Cameroun 16: 138 (1973) & in Fl. Gabon 23: 138 (1973)

NOTE. The fruits are said to be edible.

18. PLACODISCUS

Radlk. in Sitz. Akad. Wiss. München 8: 332 (1878) & in E. & P. Pf. III, 5: 321 (1895) & in E.P. IV, 165: 810 (1932)

Understorey trees, probably dioecious. Leaves pinnate; leaflets in 2–9 pairs, glabrous with exceptionally prominent minor vein reticulum. Inflorescences borne on the trunk and branches, thyrsoid, pseudo-racemose or spike-like. Flowers regular; calyx united basally to half way or more, 5-lobed; petals absent; disk circular, broad, forming a flat platform. Stamens 7–8; filaments folded in bud. Ovary deeply 3-lobed; style equalling the ovary, 3-grooved. Fruit indehiscent, shallowly or deeply lobed or of 1–several scarcely joined mericarps. Seed without aril.

A poorly known genus of ± 15 species, mostly in west tropical Africa.

* I have not seen material from **K** 2. Davies also records it from Ethiopia but I am not sure on what grounds. Vollesen does not include it in his Ethiopian account.

Leaflets in 3–5 pairs, 10–26 × 4–13 cm.; thyrse simple, 27–40 cm. long, spike-like, the branches reduced to sessile fascicles (**T** 3) · 1. *P. amaniensis*
Leaflets in 6–7 pairs, 6–12 × 2–3 cm.; thyrse branched, axis up to 27 cm. long (**T** 6) · 2. *P. pedicellatus*

1. **P. amaniensis** *Radlk.* in E.P. IV, 165: 813 (1932); T.T.C.L.: 560 (1949). Type: Tanzania, Lushoto District, Amani, *Zimmermann* in *Herb. Amani* 1477 (B†, holo., EA!, iso.)

Weak-stemmed tree ± 15 m. tall. Petiole 5–17 cm. long, woody; rhachis 4–20 cm., glabrous; petiolules swollen, channelled, 4–10 mm. long; leaflets in 3–5 pairs, oblong-elliptic or lanceolate, the uppermost pair largest, 12–26 cm. long, 4–13 cm. wide, the apex with a short acute acumen, the base unequally cuneate or rounded; both surfaces glabrous and shining, drying brown; lateral nerves in ± 10 well-marked pairs, they and the midrib with a convex line on the upper surface, very prominent beneath, the veins forming a very fine raised reticulum on both surfaces. Inflorescence unbranched, with sessile fascicles on long spikes, apparently borne on the stem, opposite or between leaves; axis 27–40 cm. long, puberulous, bearing cymose groups of 3–7 flowers subtended by subulate bracts. Calyx-tube 1.5 mm. long, lobes 2 mm. long, densely ferrugineous-pubescent; disk said to be densely pubescent. Stamen filaments pilose. Fruit unknown.

TANZANIA. Lushoto District: E. Usambara Mts., Amani, Nov. 1907, *Zimmermann* in *Herb. Amani* 1477! & 18 Dec. 1928, *Greenway* 1044! & 21 Dec. 1928, *Greenway* 1067!
DISTR. **T** 3; not known elsewhere
HAB. Rain-forest; 900 m.

NOTE. Despite a close similarity to *P. leptostachys* Radlk. this does appear to be distinct from any West African species, with broader leaves, very fine reticulation and sessile flowers. Radlkofer saw buds only; the surviving type material unfortunately has bare thyrse axes only. Both Greenway specimens mention that it is common in the Amani forest area but no further material has been seen.

2. **P. pedicellatus** *F.G. Davies* sp. nov. a congeneribus axibus secondariis elongatis pedicellisque longioribus 5–9 mm. longis differt. Typus: Tanzania, Ulanga District, Kilombero valley, Sanje logging camp, *Rodgers* 383 (K!, holo., ?DSM, iso.)

Tree ± 20 m. tall. Petiole 3.5–6 cm. long, glabrescent; leaf-rhachis 13–17 cm. long, minutely hairy; petiolules 2 mm. long, swollen, pubescent; leaflets in 6–7 pairs, drying brownish, alternate apart from the lowermost, narrowly oblong-elliptic to lanceolate, 6–12 cm. long, 2–3 cm. wide, the middle pairs slightly longer than the lowermost, abruptly acuminate, the base asymmetrically acute or rounded, both surfaces glabrous; lateral veins in ± 13 pairs, indistinct above, prominent beneath, the midrib with a raised line on the upper surface, minor vein reticulum prominent on both surfaces. Inflorescence a branched thyrse borne on the old wood; main axis 27 cm. long, branches up to 11 cm. long, pubescent; flowers single or in groups on the main or branch axes or on short minor branches; bracts 0.5–2 mm. long; pedicels 5–9 mm. long, sometimes articulated with a pedicel-like stalk 1–5 mm. long below. Calyx-lobes triangular, 1.5 mm. long, externally shortly velvety pubescent; disk pubescent. Staminodes 8 (male flowers unknown). Ovary deeply 3-lobed, 3-locular; style equalling the ovary. Fruit unknown, expected to be a group of 1–3 mericarps joined at the base only or for a greater part of the central septum. Fig. 19.

TANZANIA. Ulanga District: Kilombero valley, Sanje logging camp, 30 Dec. 1980, *Rodgers* 383!
DISTR. **T** 6; known only from the type
HAB. Rain-forest; 1220 m.

FIG. 19. *PLACODISCUS PEDICELLATUS* – **1**, leaf, × $^2/_3$; **2**, inflorescence, × $^2/_3$; **3**, female flower, × 4; **4**, female flower with part of calyx removed, × 4; **5**, transverse section of ovary (diagrammatic), × 8. All from *Rodgers* 383. Drawn by Pat Halliday.

NOTE. This new species has much of the appearance of West African members of the genus, but the long secondary axes and pedicels give the inflorescence a quite distinct aspect.

19. PANCOVIA

Willd., Sp. Pl., ed. 4, 2: 285 (1799); Radlk. in E.P. IV, 165: 799 (1932), *nom. conserv.*

[*Erioglossum* sensu G.P. 1: 396 (1862) pro parte quoad spec. ex Africa; Baker in F.T.A. 1: 420 (1868) quoad *E. cauliflorum*, *non* Blume]

Small trees or shrubs. Leaves pinnate; leaflets entire. Inflorescences axillary, terminal or from older wood. Flowers slightly zygomorphic; calyx of (4–)5 partly united sepals, the sinuses between lobes unequal; petals (3–)4, clawed, scale lobed or elaborately fringed and appendaged on the abaxial surface; disk variously semicircular or complete, variable within same species. Stamens 7–11; filaments nearly glabrous to densely hairy. Ovary 3-lobed; style shortly 3-fid at the apex. Fruit indehiscent, either 1–3 mericarps free from the base to narrowly joined for $^2/_3$ their length, or a 3-lobed drupe.

About 13 species in tropical Africa, especially in the west; one extending to temperate South Africa. The genus is one of great difficulty and no proper account will be possible until correlated open male and female flowers and fruits are available from every locality from which they are recorded. A great deal of material is only in bud. This account is unsatisfactory.

1. Male inflorescences very condensed, 1–3 cm. long, with no space between individual flowers; female flowers not condensed; leaflets (1–)3–4-jugate (**T** 3, 6–8)* ······· 1. *P. holtzii*
   Male inflorescences with individual flowers or cymules separated along the rhachis, 0.3–20.5 cm. long ······ 2
2. Leaves unknown; ultimate cymules of inflorescence shortly stalked (**K** 7, Voi) ······ 4. *P. hildebrandtii*
   Leaves present; ultimate cymules of inflorescence ± sessile ······ 3
3. Leaflets 3–5-jugate, only rarely a few leaves reduced to 2 pairs of leaflets (mainly coastal) ······ 2. *P. golungensis*
   Leaflets predominantly 2-jugate, only rarely some leaves with 3 pairs (mainly western) ······ 4
4. Leaflets narrowly elliptic-lanceolate, 5–8(–14) cm. long, 1.3–2.3(–3.8) cm. wide (poorly known species from **T** 7, Mufindi at 1450 m.) ······ 6. *P. sp. B***
   Leaflets more broadly oblanceolate, obovate or elliptic-lanceolate, 3.5–20 × 1.7–7.5 cm. (**U** 2, 4, **T** 1) ······ 3. *P. turbinata*

1. **P. holtzii** *Radlk.* in E.P. IV, 165: 802 (1932); T.T.C.L.: 560 (1949); Vollesen in Opera Bot. 59: 58 (1980). Type: Tanzania, Uzaramo District, Dar es Salaam, *Holtz* 1015 (B†, holo., M!, fragment)***

* If species 5 (*P. sp. A*, p. 62) is *P. harmsiana* it also will key here but has shorter even more closely adpressed darker brown indumentum on the buds and much longer pedicels.

** *P. sp. A* from **T** 4, known from two sterile specimens, has leaflets in 2–3 pairs.

*** There is a Holtz specimen of this in EA originally labelled '*Popowia holtzii*' without locality. It bears a number G7513 (G = German) given it by Greenway much later. There is a possibility it could be an isotype. There is no *Popowia holtzii* and it is clearly a slip.

Dioecious tree or shrub, (0.5–)4–20 m. tall, with blackish bark; young shoots ± pubescent. Leaves drying grey-green; petiole 6–14 cm. long; rhachis 3.5–12 cm. long, glabrous or minutely pubescent; petiolules 0.5–1.1 cm. long, thickened and ± wrinkled, pubescent; leaflets usually in 3–4 pairs (see note), oblong-oblanceolate, oblong or rarely obovate to oblong-elliptic, 5–22 cm. long, 2–8 cm. wide, often broadest just below apex, acuminate at the apex, the actual tip rounded or acute, cuneate at the base, coriaceous; midrib glabrous or pubescent near base, lateral nerves in 6–10 pairs. Inflorescences axillary and on older wood; male condensed, 1–3 cm. long, with no space between individual flowers, densely orange-brown velvety pubescent; female 3–3.5(–8.5) cm. long, with darker ferruginous indumentum and flowers or cymules more scattered; pedicels ± 2 mm. long; bracts 3 mm. long, 1–2 mm. wide. Male flowers: calyx-tube 5 mm. long, with unequal ovate lobes 1–3 mm. long; petals 4, yellowish white with ovate lamina and long hairy claw, 7–10 mm. long, up to 5 mm. wide, scale bilobed, 5 mm. long, apically erose and often with a long forked appendage from each lobe; disk reniform, flat, 2 mm. wide; stamens 6–11; filaments 2.4 mm. long. Female flowers: calyx-tube 2 mm. long, lobes 2–2.5 mm. long; fertilised ovary ovoid, 3 mm. long, densely pubescent; style and stigma 3 mm. long; staminodes ± 2 mm. long. Fruit greenish yellow, obtriangular, 3 cm. long, 2 cm. wide, deeply 3-lobed; endocarp acutely triangular in section. Seeds oblong, 1.7 cm. long, 7 mm. wide, finely reticulate.

subsp. **holtzii**

Leaflets acuminate, with venation beneath much more prominent and that above less evident. Fruiting material not seen.

TANZANIA. Uzaramo District: Kisarawe, Kisiju, Kiregese [Kurekese] Forest Reserve, Sept. 1953, *Semsei* 1372! & Vikindu Forest Reserve, Aug. 1953, *Paulo* 126!; Rufiji District: Kichi Hills, 22 Dec. 1976, *Vollesen* in *M.R.C.* 4259!; Iringa Districts: Kilombero, Mwanihana Forest Reserve, above Sanje, 30 Aug. 1984, *D.W. Thomas* 3633!

DISTR. **T** 6, 7, 8 (*fide* Vollesen); not known elsewhere

HAB. Coastal thicket and evergreen forest; 20–450(–750*) m.

NOTE. Undoubtedly correlated female flowers are needed. The description is taken from specimens which from geography must be assumed to be females of *P. holtzii*, e.g. Pande, 18 July 1982, *Hawthorne* 1303, with 1–3 jugate leaves, & 11 Dec. 1976, *Mhoro & Wingfield* 2410, with 2-jugate or abnormally trifoliolate leaves but apical foliage in both cases. They had been named as *P. golungensis* and this problem needs attention.

Davies mentions sparse to dense glands on undersides of the leaflets and Radlkofer makes use of this character in his keys. The translucent spots are so small, however, that they can only be seen under a microscope and I consider reference to them is confusing.

subsp. **faulknerae** *Verdc.* subsp. nov., a subsp. *holtzii* foliolis vix acuminatis, venatione supra prominentiori subtus minus evidenti differt. Typus: Tanzania, Pangani District, Bushiri, *Faulkner* 751 (K!, holo.)

Leaflets not or scarcely acuminate, with venation beneath less prominent but that above more so. Fruits as in specific description.

TANZANIA. Tanga District: Kiwanda, Sigi R., 10 Nov. 1981, *Kibuwa* 5455!; Pangani District: Bushiri, 10 Dec. 1950, *Faulkner* 751!

DISTR. **T** 3; not known elsewhere

HAB. Forest including riverine; ± 45 m.

NOTE. No flowers have been seen; until correlated flowers and fruits of both taxa are known the above arrangement is supposition.

* Possibly 900 m.

2. **P. golungensis** (*Hiern*) *Exell & Mendonça*, C.F.A. 2: 85 (1954); Exell in F.Z. 2: 525, t. 109 (1966) pro parte excl. spec. *White* 3729; Vollesen in Opera Bot. 59: 58 (1980); Beentje, K.T.S.L.: 422, fig. (1994). Type: Angola, Golungo Alto, *Welwitsch* 4516 (LISU, holo., BM!, K!, iso.)

Shrub or small tree 1.8–18 m. tall with glabrous branches; bark pale grey-brown, described as rough or smooth; twigs densely lenticellate. Leaves often drying grey-green; petiole slender, 2.5–15 cm. long, thickened at base, glabrous; rhachis 6–12 cm. long; petiolules 5–7 mm. long; leaflets in (2–)3–5 pairs, rarely 2 but probably always some with more (see note), oblanceolate to broadly elliptic-oblong, obovate or elliptic, 5.5–27+ cm. long, 1.7–7(–12) cm. wide, rounded to subacute or distinctly acuminate or rarely emarginate at the apex, cuneate at the base, glabrous. Inflorescences axillary and on the older wood, narrowly thyrsoid with single flowers or subsessile ± 3-flowered cymules spaced at intervals, ferruginous pubescent; male 2–20.5 cm. long, pedicels 0–2.5 mm. long; female mostly shorter, 2.5–5(–7) cm. long, pedicels 2–2.5 mm. long (in Flora area – 5–7 mm. in Natal). Male flowers: calyx-tube 1–2 mm. long, lobes triangular, 2–3 mm. long, velvety brown-pubescent outside; petals white or cream, obovate, the limb 2 mm. long and wide, scale bilobed, with a pair of simple or forked ligulate appendages; disk semicircular; stamens 8 with pilose filaments 7 mm. long. Female flowers: ovary 1.8 mm. long, ferruginous pubescent; style and stigma 3 mm. long. Fruit orange-yellow, with 1–3 mericarps developing, joined for up to $^{2}/_{3}$ their length, each ellipsoid, compressed, 1.5–2 cm. long, 1.1–1.5 cm. wide, up to 1.1 cm. thick, sometimes only one developing then other 2 remaining at its base, ± 6 mm. long; pedicels ± 3 mm. long. Seeds pale brown or dark, compressed ellipsoid, 1.2–1.8 cm. long, 0.8–1.3 cm. wide, up to 8 mm. thick.

KENYA. Kwale District: Shimba Hills, Mwele Mdogo Forest, 5 Feb. 1953, *Drummond & Hemsley* 1130!; Kilifi District: Mida, Apr. 1938, *Dale* in *F.D.* 3875!; Lamu District: Witu Forest, 15 Nov. 1988, *S.A. Robertson & Luke* 5478!

TANZANIA. Morogoro District: Turiani, Nov. 1954, *Semsei* 1851!; Kilwa District: 10 km. NNW. of Kingupira, 3 Dec. 1976, *Vollesen* in *M.R.C.* 4181! & 12 km. NNW. of Kingupira, 25 Oct. 1976, *Vollesen* in *M.R.C.* 4069!

DISTR. **K** 7; **T** 6, 8; Angola, Mozambique and South Africa (Kwazulu-Natal)

HAB. Moist evergreen lowland forest and riverine forest; 0–500(–900) m. (see note)

SYN. *Aphania golungensis* Hiern, Cat. Afr. Pl. Welw. 1: 169 (1896)
[*Pancovia turbinata* sensu Radlk. in E.P. IV, 165: 804 (1932) pro parte, *non* Radlk. sensu stricto]
[*P. hildebrandtii* sensu Beentje, K.T.S.L.: 422 (1994) pro parte, *non* Gilg]
P. sp. aff. ugandensis sensu F.G. Davies ined.; Beentje, K.T.S.L.: 421 (1994)

NOTE. A series of specimens from the Kenya coast with leaves having only two pairs of leaflets (*S.A. Robertson & Luke* 5964, Kwale District, Gongoni Forest Reserve, 12 Nov. 1989, shrub to 1.5 m. & *S.A. Robertson & Luke* 6316, same locality, 1 June 1990, shrub to 75 cm. & *S.A. Robertson & Luke* 2803, Shimba Hills, Mkongani West, 25 Mar. 1991, shrub 1 m.) have been named *P. sp.* near *P. ugandensis* F.G. Davies - a manuscript name proposed for the species here treated as *P. turbinata* Radlk. I have left these specimens as forms of *P. golungensis* but this is more evidence that the latter might be better treated as a synonym of *P. turbinata* as indeed was done by Radlkofer. *Faden et al.* 71/138 (Teita District, Mt. Kasigau, pipeline from Rukanga, 6 Feb. 1971, 1250–1450 m.) is sterile and all the leaves (young) appear to be 2-jugate but it matches other **K** 7 material in facies. *Mwasumbi & B.J. Harris* in *D.S.M.* 2685 (Morogoro District, Kimboza Forest Reserve, 10 Sept. 1972) also appears to have only 2-jugate leaves but more adequate material is needed – leaves from branch tips may not be typical. *Drummond & Hemsley* 1130 is described as a small shrub to 1.8 m. with spreading branches, some also prostrate.

Only the Natal populations are adequately known and these reach into S. Mozambique; the leaves are (2–)3–5-jugate, the leaflets rounded to acuminate at the apex; the male inflorescences are ± 3 cm. long and in one specimen rather similar to those of *P. holtzii*; the female inflorescences are 5–7 cm. long, with pedicels 3–4.5(–6) mm. long; style and stigma 3 mm. long. The **T** 6 and 8 populations have leaflets drying paler grey-green, rounded, subacute or emarginate at the apex; male inflorescences up to 18 cm. long, but females inadequately known; fruiting pedicels ± 7 mm. long. The Kenya coast populations have much more acuminate leaflets, female inflorescences with pedicels only 2–2.5 mm. long. Unfortunately even less is known about the population in the type area in Angola but as matters stand there seems no adequate reason to recognise distinct taxa. The occurrence of *P. holtzii* between the populations of *P. golungensis* with the females at least being difficult to separate is unsatisfactory. The genus is in great need of field study.

3. **P. turbinata** Radlk. in Sitz. Akad. Wiss. München 8: 270 (1878) & in E.P. IV, 165: 804 (1932) pro parte; F.W.T.A., ed. 2, 1: 718 (1958). Type: Guinea, *Heudelot* 869 (P, holo., K!, iso.)

Tree 4–15(–25 *fide* Troupin) m. tall or shrub in West Africa, sometimes said to be sprawling or with gnarled trunk; crown dense; bark brown or red-brown, thin, smooth, horizontally striated; young shoots ferruginous pubescent; slash (in Uganda) granular, pale brown or off-white, vaguely mottled with orange-brown. Leaves drying grey-green, with (1–)2(–3) pairs of leaflets or abnormally pseudo-trifoliolate; petiole dark brown, 1.5–10 cm. long, puberulous or glabrescent; rhachis 1–5 cm. long; petiolules ± thickened, 2–10 mm. long; leaflets equal or upper pair up to twice as large, oblanceolate-obovate, elliptic or lanceolate, 3.5–15(–20) cm. long, 1.7–6.3(–7.5) cm. wide, rounded to mostly shortly acuminate at the apex, the tip rounded or acute, cuneate at the base, ± coriaceous, glabrous; lateral nerves in 8–12 pairs. Inflorescences single or 2–3 closely placed, axillary or at nodes when leaves have fallen, 0.3–5 cm. long, densely dark ferruginous pubescent; flowers spaced, either single or in 3-flowered clusters; pedicels ± 1 mm. long, but buds subsessile; bracts ovate, 1.5 mm. long and wide. Male flowers: calyx-tube ± 1 mm. long, with triangular lobes 1.5–2 mm. long; petals 5, white, cream or greenish yellow, ± 2–2.5 mm. long, the claw ± 1 mm. long, scale half as long as the limb, entire or bifid with lobed ciliate appendage; disk circular or unilateral, reniform; stamens 7–8(–9); filaments 3.5 mm. long (but probably longer in adult flowers), pilose below. Female flowers: ovary deeply 3-lobed; stigma capitate; staminodes 8, very reduced. Fruits not seen (ellipsoid, 1 cm. long ex icon. Troupin). Fig. 20.

UGANDA. Kigezi District: Ishasha Gorge, 10 Feb. 1945, *Greenway & Eggeling* 7098!; Masaka District: Sese Is., Funve I., 19 June 1950, *G.H.S. Wood* Y63!; Mengo District: Kasa Forest, near Mityana, 31 Jan. 1950, *Dawkins* 506!

TANZANIA. Bukoba District: Kiau [Kiao] I., *Gillman* 391!; Ngara District: Nyamiruma R., July 1953, *Eggeling* 6665!; Mwanza District: Rubondo I., Lukaya, 10 May 1974, *Ludanga* in *M.R.C.* 2039!

DISTR. **U** 2, 4; **T** 1; Ivory Coast, Cameroon, Equatorial Guinea and Rwanda

HAB. Evergreen rain-forest, including riverine; 1150–1500 m.

SYN. *P. sp.* near *P. turbinata* Radlk. sensu I.T.U., ed. 2: 382 (1952); Hamilton, Field Guide Uganda For. Trees: 225, fig. 417 (1981)

P. sp. A sensu Troupin, Fl. Rwanda 2: 306, fig. 97/1 (1983)

NOTE. Davies had intended to describe this as a new species; it had for long been indicated as such. It does not seem to me to be sufficiently separable from the West African material. Radlkofer eventually also included *Aphania golungensis* in the synonymy but that at least usually has more leaflets (the type has 4–5-jugate leaflets); the material available is hopelessly inadequate to be sure they are only variants of one species and are best kept separate until correlated fruits, male and female flowers and leaf spectra are available. Davies mentions the Uganda material has the leaflets sometimes glandular beneath and Radlkofer excludes *P. turbinata* from those which do in his key but I think this is a very dubious character. Fouilloy & Hallé (in Fl. Cameroun 16: 118 (1973) & in Fl. Gabon 23: 118 (1973)) put *P. turbinata* in the synonymy of *P. pedicellaris* Radlk. & Gilg, a name published in 1897.

4. **P. hildebrandtii** Gilg in E.J. 24: 304 (1897); Radlk. in E.P. IV, 615: 807 (1932); K.T.S.: 516 (1961); Beentje, K.T.S.L.: 422 (1994) pro parte. Type: Kenya, Teita District, Voi R., *Hildebrandt* 2491 (B†, holo., M!, fragment)

FIG. 20. *PANCOVIA TURBINATA* — **1**, flowering branch, × $^{2}/_{3}$; **2**, male flower, × 6; **3**, male flower with one sepal and two petals removed, × 6; **4**, petal from male flower, × 10; **5**, female flower with one sepal removed, × 6; **6**, petal from female flower, × 10. 1, 5, 6, from *Katende & Lye* 353; 2, 3, from *Dawkins* 506; 4, after Troupin. Drawn by Pat Halliday.

Scandent shrub, the floriferous branchlets 2.5–3 cm. in diameter. Leaves unknown. Inflorescences 2–6-fasciculate, borne at thickened leafless nodes, paniculate or pseudoracemose, 5–7 cm. long, the rhachis pale ferruginous tomentose with very dense stiff sharply pointed hairs; individual cymules subcapitate, 3–5-flowered with stalks 5–8 mm. long; pedicels 2 mm. long; bracts ± 2 mm. long. Male flowers: calyx-lobes 5, joined to middle, elliptic, 5 mm. long, 1.8(–4) mm. wide (see note), broadly imbricate, subacute, densely pale ferruginous tomentose outside; petals 4, slightly exceeding the sepals (or only 2.5 × 1.5 mm. – see note) the lower $^1/_3$ comprising a thick claw, blade membranous, the margin crenate-lobulate, scale adnate to petal above the claw, with a variable deeply lobed or lobulate crest slightly exceeding the margin of the scale; stamens 7 (3 mm. long in fragment) with elongate pilose filaments slightly exceeding the petals; anthers large (1.5 mm. long in fragment), glabrous, dorsally affixed, the connective drying black, dilated; disk fleshy, very excentric, lobulate, glabrous; rudimentary ovary absent. Female flowers and fruits unknown.

KENYA. Teita District: Voi R. at Voi, 6 Feb. 1877, *Hildebrandt* 2491!
DISTR. **K** 7; known only from the type
HAB. Presumably riverine forest; probably ± 550 m.

NOTE. Radlkofer suspected this was not a *Pancovia* but Davies considered it a synonym of *P. golungensis*. I have re-examined the Munich fragment. The character, inflorescence elements with stalks 5–8 mm. long, does not fit any other material seen (admittedly not completely mature) in which they are all sessile and there are more distinct bracteoles in *P. golungensis*. I do not think they are conspecific. Radlkofer gives the calyx-lobes as 5 × 4 mm. and the petals as slightly exceeding the sepals whereas in the fragment seen the calyx-lobes are 5 × 1.8 mm. and the petals only 3 × 1.8 mm. - a difference in maturity is indicated. Although some coastal specimens of *P. golungensis* do have very short cymule-stalks to 2 mm. (e.g. *S.A. Robertson & Luke* 5990, Kwale District, Maluganji Forest Reserve, 14 Nov. 1989 & *S.A. Robertson & Luke* 6076, Kilifi District, Kaya Kivara, 23 Nov. 1989) they are not the same as *P. hildebrandtii*. The type fragments have a different velvety orange-fulvous indumentum and different sepals apart from longer cymule stalks. I have been unable to correlate the material with any other species and assume that it is, despite the fact the area has been fairly well collected, something never refound. [B.V.]

5. **P. sp. A**

Tree 5–6 m. tall; bark whitish grey, smooth; innovations ferruginous hairy, otherwise glabrous. Leaves with 2–3 pairs of leaflets; petiole 2.2–9 cm.; rhachis 2–10 cm.; petiolules 6 mm.; leaflets elliptic, somewhat acuminate, the base cuneate, the uppermost pair slightly larger than the lower, 7.5–20 × 2–7 cm.

TANZANIA. Kigoma District: Kabogo, 21 June 1963, *Toyoshima* in *Kyoto University Expedition* 577!; Mpanda District: Lukandamila, May 1972, *Kakeya* 111!
DISTR. **T** 4; see note
HAB. Riverine forest; 1150–1600 m.

NOTE. These sterile specimens may be *P. harmsiana* Gilg but the material is inadequate. *P. harmsiana* Gilg is a species from Zaire. It has densely clustered inflorescences similar to those of *P. holtzii*; the fruits are fleshy mericarps, ovoid or subspherical, 12 mm. across.

6. **P. sp. B**

Tree 10–15 m. tall; bark brown, smooth; slash pale brown, granular, slightly layered. Leaflets 1–2-jugate, narrowly elliptic-lanceolate, 5–8(–14 in Malawi) cm. long, 1.3–2.3(–3.8 in Malawi) cm. wide, acuminate at the apex, cuneate at the base. Young male inflorescences 1.2 cm. long, but flower clusters will be scattered at maturity (one inflorescence in Malawi specimen reduced to a single flower sessile in axil of fallen leaf; sepals triangular, 1.5 mm. long; filaments 4 mm. long).

TANZANIA. Iringa District: Mufindi, Lulando Forest, 19 Apr. 1987, *Lovett & Congdon* 2024!
DISTR. **T** 7; N. Malawi
HAB. Closed canopy evergreen forest; 1450 m.

SYN. *P. sp. 1* sensu F.F.N.R.: 225 (1962)
[*P. golungensis* sensu Exell in F.Z. 2: 528, t. 109/1 (1966) pro parte, *non* (Hiern) Exell & Mendonça sensu stricto]

20. CHYTRANTHUS

Hook.f., G.P. 1: 403 (1862); Radlk. in E.P. IV, 165: 782 (1932)

Glossolepis Gilg in E.J. 24: 299 (1897); Radlk. in E.P. IV, 165: 778 (1932)

Small trees or shrubs, often of palm-like habit with unbranched trunks. Leaves pinnate, large; leaflets many, glabrous, some with epidermal glands. Inflorescence arising variously from roots, trunk or branches, pseudoracemose or thyrsoid; plants probably usually dioecious. Flowers slightly irregular with 4 petals crowded to one side and the stamens or ovary at the opposite side; calyx unequally 5-lobed, united into a cup to half-way or more; petals 4, oblong, clawed, with a tongue-like scale up to $^{2}/_{3}$ as long as the limb and sometimes bearing a simple or lobed appendage on its abaxial side; disk asymmetrical, lobed. Stamens 7–8(–15); filaments conspicuously hairy; anthers pilose or glabrous. Ovary lobed, 3–several-locular; style entire. Fruit an indehiscent lobed or ribbed drupe. Seed 1 per locule without arils.

A genus of 30 or more species in Africa.

1. Inflorescence-axes with solitary flowers or cymules subtended by conspicuous subulate bracts up to 2 cm. long (**T** 6) . . . . 4. *C. longibracteatus*
   Inflorescences with elements not supported by visible bracts or bracts only up to 2 mm. long 4
2. Flower buds drying dark purple; leaves with 2–6 pairs of leaflets; ovary and fruit 5–6-locular (**T** 1, 7) . . . . 2. *C. atroviolaceus*
   Flower buds drying orange-brown but young leaf-buds sometimes purple-hairy; leaves with 3–12 pairs of leaflets; ovary and fruit 3-locular 3
3. Inflorescences from base of trunk just below or just above ground-level; a subshrub 0.2–2 m. tall; leaves up to 50 cm. long, with 3–6 pairs of leaflets up to 23 cm. long; fruits fleshy, trigonous, ± 5 cm. long and wide, finely pubescent, the fruiting inflorescence usually lying on the ground (**K** 7; **T** 6, 8; **Z**) . . . . 1. *C. prieurianus*
   Inflorescences on trunk and branches (if any) probably sometimes near ground-level; mostly a small tree but can be subshrubby, 1.2–6(–15) m. tall; leaves 1 m. long, with (5–)7–12 pairs of leaflets up to 45 cm. long; mature fruits unknown but probably always hanging from the trunk and branches, densely hairy at least in the young state (**K** 7; **T** 3) . . . . 3. *C. obliquinervis*

1. **C. prieurianus** *Baill.* in Adansonia 11: 241 (1874); Radlk. in E.P. IV, 165: 793 (1932); F.W.T.A., ed. 2, 1: 718 (1958); Hallé in Bull. Mus. Hist. Nat. Paris, sér. 3, 127: 11, fig. 1–4 (1973). Type: from Zanzibar, cultivated in Paris, July 1866 (P, holo. & living original specimen in glasshouse at Paris)

Slender single-stemmed dioecious shrub, (0.2–) 1–2 m. tall, from a thick starchy root; bark slightly rough with raised lenticels which when young are covered with quartered or stellate cork cell masses. Leaves up to 50 cm. long, with 3–6 pairs of leaflets; petiole 5–15 cm. long, ± woody, brownish puberulous; rhachis 9–19 cm. long; petiolules ± thickened, 6–8 mm. long, brown-pubescent; leaflets opposite or alternate, the leaves sometimes appearing pseudo-imparipinnate, oblanceolate, obovate-lanceolate or elliptic-oblong, 4.5–23(–?32) cm. long, 2.9–8(–?11.5) cm. wide, acuminate at the apex, the actual tip ± obtuse or slightly acute, cuneate to truncate or very slightly subcordate at the base, glabrous save occasionally for few hairs on midrib beneath. Inflorescences pseudoracemose, borne on the stems just above or below ground-level, 4.5–30 cm. long, brown-puberulous; peduncle 7.5–25 cm. long; individual cymules numerous, up to ± 50, each 3–6-flowered. Flowers sweet-scented; buds grey-tomentose; pedicels up to 1.3(–2) mm. long; calyx pale green, campanulate to cylindric, 3–5.5 mm. long, shortly densely pilose, lobes unequal, 1–1.5 mm. long, 0.75–1.75 mm. wide; petals 4–5, cream, narrowly clawed, the claw linear (0.5–) 1–2.5 mm. long, limb elliptic-oblong, (2–) 4–5.5 mm. long, (0.5–) 2 mm. wide, lobulate at the apex, margin crenulate, scale very variable, up to $^1/_3$ the length of limb, adnate at the base or sometimes throughout, entire or bilobed at the apex, lobes simple, triangular, 0.5 m. long, a narrow horn-like crest sometimes present; disk semilunar, unilateral. Stamens 7–9; filaments 3–5 mm. long, pubescent. Ovary ellipsoid, 2.5 mm. long, 2 mm. wide, 3-angled, 3-locular; style conical, 2.5 mm. long, 1 mm. wide, the stigmatose apex clavate and sulcate. Fruiting inflorescence mostly prostrate on ground; fruit fleshy, orange, 3-angled-subglobose or obtetrahedral, 3-lobed, (2.7–) 3.6–5 cm. long and wide, acuminate at apex, rounded at base, the lobe margins acute, minutely pubescent. Seed ellipsoid or subdiscoid, 1.5–1.8 cm. long, 1–1.5 cm. wide, 4–6 mm. thick, distinctly compressed, pale brown; hilum linear, 9 mm. long; testa scurfy-reticulate when dry. Fig. 21.

KENYA. Kwale District: Shimoni, 20 Aug. 1953, *Drummond & Hemsley* 3926! & Kaya Kinondo, 16 July 1987, *S.A. Robertson & Luke* 4911!; Mombasa District: Bamburi, 19 Apr. 1996, *S.A. Robertson & Baer* 7191!

TANZANIA. Morogoro, near Chief Conservator's house, Jan. 1954, *Eggeling* 6778!; Ulanga District: near Ifakara, Ifangera, 1 July 1959, *Haerdi* 264/0; Kilwa District: Selous Game Reserve, Nandembo, 19 Dec. 1976, *Vollesen* in *M.R.C.* 4241!; Zanzibar, Fumba, 9 Feb. 1930, *Vaughan* 1212!

DISTR. **K** 7; **T** 6, 8; **Z**; not known elsewhere

HAB. Evergreen forest, ground-water forest, forest on coral rag; 0–600 m.

SYN. *C. sacleuxii* Pierre & Sacleux in Bull. Mus. Nat. Hist. Paris, sér. 2, 6: 110 (1934). Type: Zanzibar, plant grown in Paris from seeds sent from Kombeni, *Sacleux* 521 (P!, holo.)

C. sacleuxii Pierre & Sacleux subsp. *longiflorus* Verdc. in K.B. 10: 604 (1956); K.T.S.: 508 (1961); Haerdi in Acta Trop., suppl. 8: 124 (1964). Type: Tanzania, Morogoro District, Lusunguru Forest Reserve, near Mtibwa sawmills, *Drummond & Hemsley* 1937 (K!, holo., EA!, K!, iso.)

C. prieurianus Baill. subsp. *longiflorus* (Verdc.) Hallé in Bull. Mus. Hist. Nat. Paris, sér. 3, 127: 19, fig. 1/8', 3/5–6, 3/12–14, 3/24, 3/29, 4/7–9 (1973); Vollesen in Opera Bot. 59: 58 (1980); Beentje, K.T.S.L.: 415 (1994)

NOTE. The type of *C. prieurianus* was originally thought to have been grown from Senegambian material but Hallé showed that it is the same as *C. sacleuxii* described much later by Sacleux from Zanzibar, although I am not certain how he knew the seeds came from Zanzibar rather than the Zanguebar coast of East Africa. It had even been thought that the original material had been collected in French Guiana by Leprieur but Baillon thought there must have been confusion with Leprieur's Senegal collections. Hallé kept up subsp. *longiflorus* mentioning that the epidermal cellular reticulation had 1.5–2 areoles per 0.1 mm. as against 2–3 in the type, also that the scale of the petal was very variable and sometimes adnate without a free lobe. These do not seem to be good reasons for maintaining the subspecies. No recent specimen has been seen from Zanzibar.

FIG. 21. *CHYTRANTHUS PRIEURIANUS* — **1**, habit, much reduced; **2**, foliage, × $^2/_3$; **3**, inflorescence, × $^2/_3$; **4**, male flower, × 4; **5**, petal from male flower, × 6; **6**, female flower with part of calyx and 2 petals removed, × 4; **7**, **8**, petals from female flower, × 6; **9**, female flower with part of calyx removed, × 4; **10**, fruits, × $^2/_3$; **11**, transverse section of fruit, diagrammatic. 1–3, from *Drummond & Hemsley* 3926 & 1937; 4–9, from *Haerdi* 264/0; 10, from photograph of *S.A. Robertson & Baer* 7191 & after fig. of *Vaughan* 1212; 11, after fig. of *Vaughan* 1212. Drawn by Pat Halliday.

2. **C. atroviolaceus** *Hutch. & Dalziel*, F.W.T.A. 1: 504 (1928) & in K.B. 1929: 26 (1929); Radlk. in E.P. IV, 165: 786 (1932); F.W.T.A., ed. 2, 1: 717 (1958); Fouilloy & Hallé in Fl. Cameroun 16: 102, t. 24 (1973) & in Fl. Gabon 23: 102, t. 24 (1973). Type: Nigeria, Oban, *Talbot* 1596 (K!, holo., BM!, iso.)

Tree 5–18 m. tall. Leaf-rhachis 0.4–1 m. long, ribbed, brown-pubescent or glabrous; petiole 20–50 cm. long, 7 mm. diameter; petiolules 0.7–1.5 cm. long; leaflets (2–)4–6-jugate, elliptic or oblong, 14–42(–46) cm. long, 6.5–17 cm. wide, shortly acuminate to caudate-acuminate at the apex, asymmetrically cuneate or, especially in the lower leaflets, truncate at the base, glabrous and shiny above, minutely pubescent or glabrous beneath; lateral veins in 15–25 pairs, prominent beneath. Inflorescences purplish, borne on the trunk at base or up to 50 cm. from ground, thyrsoid, (6–)10–15(–35) cm. long; buds purple; pedicels 2 mm. long. Calyx purple*, 8–10 mm. long, 2-lipped, densely shortly velvety pubescent; petals 4–6, obovate-spathulate or oblanceolate, 7.4–9.5 mm. long, including a 1.2 mm. claw, 3 mm. wide, scale 5 mm. long, emarginate, sometimes with a bifid appendage arising just below the apex; disk asymmetric, glabrous. Stamens 7–10, the basal $^{2}/_{3}$ of the filaments hairy; anthers glabrous. Ovary 5–6-lobed, 5–6-locular, bristly with purple hairs; rudimentary ovary absent. Fruit ± star-shaped, pyriform in outline, 7–13.3 cm. wide, 4–6-lobed or winged, densely rough brown or purple spreading hairy. Seeds brown, subovate in outline, very compressed, 1.9–3.2 cm. long, 1.1–1.8 cm. wide, 6 mm. thick.

UGANDA. Bunyoro District: Budongo Forest, *Plumptre* 1! & 2–3 Aug. 1996, *Plumptre*! & 9 Apr. 1995, *Katende* & Block 1A, 22 Mar. 1997, *Reynolds & Tinka John*!

TANZANIA. Bukoba District: Kikuru Forest, Sept.–Oct. 1935, *Gillman* 447!

DISTR. **U** 2; **T** 1; Sierra Leone, Ivory Coast, Ghana, Nigeria, Cameroon, Central African Republic and Zaire

HAB. Evergreen forest, sometimes on periodically inundated grey clay loams; 1100–1300 m.

SYN. *C. brunneotomentosus* Radlk. in E.P. IV, 165: 787 (1932); F.W.T.A., ed. 2, 1: 718 (1958); Hauman in F.C.B. 9: 353, fig. 7 (1960). Types: N. Cameroon, Bangwe, *Conrau* 94 (B†, syn., M!, fragment) & 172 (B†, syn.)

C. macrophyllus Gilg var. *obanensis* Baker f. in Cat. Pl. Talbot S. Nig.: 19 (1913). Type as for *C. atroviolaceus*

C. sp. nov. sensu Verdc. in K.B. 10: 604 (1956)

NOTE. The Ugandan material has large fruits, 11.5–13.3 cm. wide, and glabrous leaves, leaf-rhachises, etc., but until more is known about the variation they seem best placed here. A specimen *Cribb et al.* 11155 from **T** 6 (Ulanga District, Sali, Ngongo Mt., 23 Jan. 1979) was referred here by Davies but the fruit is rounded.

Radlkofer gave petals 6 and rudimentary ovary 6-locular in the description of *C. atroviolaceus.*

3. **C. obliquinervis** *Engl.*, V.E. 3(2): 274 (1921); Radlk. in E.P. IV, 165: 797 (1932); T.T.C.L.: 556 (1949); Verdc. in K.B. 10: 603 (1956); K.T.S.: 508 (1961); Beentje, K.T.S.L.: 415, fig. & map (1994). Type: Tanzania, E. Usambara Mts., Derema [Nderema], *Scheffler* 185 (B†, holo., BM!, M!, W!, iso.)

Dioecious or ?monoecious tree or suffrutex, 1.2–6(–15) m. tall, mostly unbranched with apical rosette-like crown of very large leaves; bark grey-brown, smooth or somewhat wrinkled. Leaves drooping and purplish when young, with 7–12 pairs of leaflets; leaf-buds purplish; petiole 1–3 cm. long, ± woody, pubescent; rhachis 35–100 cm. long; petiolules swollen, ± 1 cm. long; leaflets opposite or alternate, elliptic, oblanceolate or oblong-lanceolate, (3–)5.5–45 cm.

* Vigne says red - perhaps only purple on drying.

long, (1.5–)4–14 cm. wide, shortly acuminate at the apex, rounded to cuneate at the base, coriaceous, glabrous save for scattered long hairs on the venation beneath; midrib sunken on upper surface. Inflorescences clusters of thyrses borne on the trunk and branches, pinkish green, 5–15 cm. long, the male and female possibly sometimes on the same plant; buds cream. Calyx pale green velvety, somewhat 2-lipped, tube 3–4 mm. long, lobes very unequal, 2.5–4 mm. long; petals white or yellow-green, oblong, 4–5 mm. long, emarginate or lobed, scale 3.5 mm. long, bent forward but not hooded at the tip, bearing a short simple appendage, hairy at the junction with the petal limb; disk reniform, deeply lobed opposite the petals. Stamens 8; filaments and anthers pilose. Ovary 3-lobed, 3-locular. Mature fruit unknown but very young ones are velvety tomentose and with longer white hairs as well.

KENYA. Kwale District: Msambweni, Buda Mafisini Forest, 21 Jan. 1964, *Verdcourt* 3955! & Shimba Hills, Mwele Mdogo Forest, 5 Feb. 1953, *Drummond & Hemsley* 1120!; Tana River District: E. of Tana R., Hewani [Haiwani] Forest, 13 Feb. 1973, *Katende* 1761!

TANZANIA. Lushoto District: Muheza–Amani road, 16 Apr. 1968, *Renvoize & Abdallah* 1514! & Amani Forest Reserve, Nenguruwe Forest, 31 Nov. 1986, *Ruffo & Mmari* 2027! & Amani, above Dodwe R., 18 Nov. 1947, *Brenan & Greenway* 8336!

DISTR. **K** 7; **T** 3, not known elsewhere

HAB. Lowland evergreen forest on sand and coral and evergreen rain-forest with *Milicia* etc.; 5–900 m.

NOTE. *Verdcourt* 3955 has both male and female inflorescences (detached) which may or may not have come from one individual. Unfortunately no note was made at the time.

4. **C. longibracteatus** *F.G. Davies* sp. nov., *C. obliquinervi* Engl. similis, floribus vel cymulis longibracteatis, bracteis usque 2 cm. longis, petalis majoribus, antherodiis glabris differt. Typus: Tanzania, Morogoro District, Nguru Mts., Turiani, Mhonda Mission, *Semsei* 1402 (K!, holo., EA, iso.)

Small dioecious tree; bark grey. Leaves with petiole 6 cm. long, 1 cm. wide at base, woody, ribbed, shortly pubescent in the grooves; rhachis up to 1 m. long, tapering, angular; leaflets crowded, ± 12-jugate, the uppermost oblong-oblanceolate or oblong, 20–33 cm. long, 5–8 cm. wide, the lowermost elliptic, 7–8 cm. long, 3–5 cm. wide, obtuse to acute or shortly acuminate at the apex, truncate at the base, glabrous; lateral veins in ± 8 pairs in small leaflets, ± 20 in upper larger leaflets, prominent. Inflorescences borne on trunk near ground and lower parts of the branches, 10–25 cm. long, at first brownish velvety; male not known; female flowers borne singly or in pairs subtended by a subulate bract up to 2 cm. long. Calyx 2-lipped with 2 triangular sepals 5 mm. long, separated by a sinus of 2 mm. and 3 sepals 4 mm. long, separated by sinuses of 1.2 mm., increasing slightly in fruit, densely pubescent; petals 4, drying reddish, oblong, 8 mm. long, 2 mm. wide with claw 0.7 mm. long; scattered hairs present, especially at the junction of the scale, scale ligulate, 4 mm. long, slightly narrower than petal-limb, hooded at the apex, with a shallowly lobed appendage; disk reniform. Staminodes with pilose filaments and glabrous anthers. Ovary triangular. Immature fruit ovoid, 2.5 cm. long, 3-ribbed, 3-locular, orange-brown velvety tomentose and with longer hairs; pedicels 6 mm. long.

TANZANIA. Morogoro District: Nguru Mts., Turiani, Mhondo Mission, Nov. 1953, *Semsei* 1402!

DISTR. **T** 6; known only from the type

HAB. Presumably forest; ± 1000 m.

NOTE. This species has foliage identical to some material of *C. obliquinervis* but no such obvious bracts have been seen in either male or female inflorescences of that species.

21. DEINBOLLIA

Schumach., Beskr. Guin. Pl.: 242 (1827); Radlk. in E.P. IV, 165: 669 (1932)

Trees, mostly (and always in the Flora area) of palm-like habit with a terminal cluster of leaves and flowers. Leaves pinnate; leaflets entire. Flowers regular, in usually extensive panicles; sepals 5, markedly imbricate, free, glabrous to densely hairy; petals 5, ciliate, scale tongue-shaped, shorter than petal-limb; disk cup-shaped, glabrous (in Flora area) or hairy, sometimes a double ring. Stamens 9–30, with long-pilose filaments; anthers rather small, lanceolate. Ovary deeply 3-lobed; style exceeding ovary, 3-furrowed, ± persistent in fruit. Fruit of 1–3 globose to ovoid indehiscent mericarps but usually 1(–2) failing to develop and persisting at base of mature one(s).

A genus of ± 40 species in Africa (particularly in the West) and Madagascar; the species are mostly difficult. The leaves subtending the inflorescence are smaller and of different appearance from mature leaves; descriptions are mostly of the latter.

1. Leaflets ± densely pubescent beneath or on both surfaces · · · 2
   Leaflets glabrous or with scattered hairs on venation beneath · · · 3
2. Calyx densely dark ferruginous pubescent; leaflets truncate to slightly subcordate at the base, acute at the apex (**U** 4; **T** 1, 4) · · · 1. *D. fulvo-tomentella*
   Calyx much paler pubescent, yellow-brown to cream; leaflets distinctly subcordate at the base, rounded at the apex (**T** 3) · · · 3. *D. borbonica* forma *subcordata*
3. Calyx glabrous · · · 2. *D. kilimandscharica*
   Calyx densely velvety tomentose to sparsely puberulous · · · 4
4. Indumentum of calyx dark ferruginous; plant of higher altitudes (900–1950 m.) · · · 2. *D. kilimandscharica* var. *adusta*
   Indumentum of calyx pale yellowish brown or cream; plant of lower altitudes (0–1050 m.) · · · 3. *D. borbonica* forma *glabrata*

1. **D. fulvo-tomentella** *Baker f.* in J.L.S. 37: 137 (1905); Radlk. in E.P. IV, 165: 688 (1932); I.T.U., ed. 2: 378 (1952); Hamilton, Field Guide Uganda For. Trees: 223 (1981). Type: Uganda, Mengo District, Lake Victoria, Bwema [Wema] I., *Bagshawe* 592 (BM!, holo.)

Slender monoecious tree or shrub, 1.2–9 m. tall*; bark grey or pale brown, thin, smooth; slash brown. Leaves up to 0.9–1.2 m. long; petiole 6–40 cm. long; rhachis up to 70 cm. long; petiolules 0–5 mm. long; leaflets in 5–10 pairs, alternate, oblanceolate to oblong, oblong-lanceolate or elliptic, 6.5–40 cm. long, 2.7–11.8 cm. wide, acute at the apex, truncate or slightly subcordate at the base, densely yellow-brown pubescent on both surfaces. Inflorescences extensive, up to 1 m. long; major axes up to 40 cm. long, with secondary axes ± 12 cm. long, densely dark ferruginous pubescent (golden pubescent in life); buds similarly pubescent; flowers fragrant, single or in clusters on the axes; pedicels 2–5 mm. long, enlarging in fruit to 2 cm. Sepals elliptic, 4 mm. long, 3–3.5 mm. wide, silky dark brown-pubescent outside; petals white, elliptic, 6–7 mm. long, pubescent outside and ciliate. Fruit golden yellow, depressed globose, of 1–2 mature mericarps, 1.5–1.8 cm. long, 1.7–2.5 cm. wide, with fairly dense spreading pubescence.

* A reference to 60 ft. (18.5 m.) must I think be erroneous.

UGANDA. Masaka District: Sese Is., Bukasa I., Dec. 1922, *Maitland* 375!; Mengo District: 16 km. on Entebbe road, Kajansi Forest, Feb. 1938, *Chandler* 2164! & Entebbe, Kitabi [Kitabe], Oct. 1935, *Chandler* 1437!

TANZANIA. Bukoba District: Kaujuma, 26 Oct. 1948, *Ford* 771!; Mpanda District: Kungwe Mts., Kasiha [Kasieha] R., 20 July 1959, *Newbould & Harley* 4482! & 20 km. N. of Kasoge, Belengi, 2 Aug. 1959, *Harley* 9125!

DISTR. **U** 4; **T** 1, 4; Cameroon, Gabon, Central African Republic, Zaire, ?Sudan and Angola

HAB. Evergreen forest, including lakeshore and riverine forest; 750–1350 m.

SYN. *D. molliuscula* Radlk. in Z.A.E.: 477 (1912) & in E.P. IV, 165: 678 (1932); F.P.N.A. 1: 521 (1948); Hauman in F.C.B. 9: 334 (1960); Fouilloy & Hallé in Fl. Cameroun 16: 63 (1973) & in Fl. Gabon 23: 63 (1973). Type: Zaire, Lake Edward, *Mildbraed* 1961 (B†, holo., BM!, sketch)

NOTE. It has been suggested this is the same as *D. pinnata* Schumach. & Thonn. but that has smaller leaflets and different pubescence; although close it is not identical.

2. **D. kilimandscharica** *Taub.* in P.O.A. C: 250 (1895); Radlk. in E.P. IV, 165: 682 (1932); F.P.N.A. 1: 521 (1948); T.T.C.L.: 556 (1949); I.T.U., ed. 2: 378 (1952); Hauman in F.C.B. 9: 336 (1960); K.T.S.: 509, fig. 92 (1961); Vollesen in Fl. Ethiopia 3: 502, fig. 125.9 (1990); Beentje, K.T.S.L.: 416, fig., map (1994). Type: Tanzania, Kilimanjaro, Marangu, *Volkens* 1471 (B†, holo., K!, iso.)

Small tree or shrub 1.2–12 m. tall, with simple trunk; bark brownish grey. Leaves with petiole 5–20 cm. long and rhachis 15–40 cm. long, both mottled brown or green and straw-coloured towards the base when dry, adpressed-pubescent when young; petiolules 3–10 mm. long, at first pubescent; leaflets in (3–)5–7 pairs, subopposite or alternate, oblanceolate to oblong-elliptic, 4–27(–30) cm. long, 1.9–10.2(–16) cm. wide, acuminate (with actual tip rounded) or broadly rounded to truncate at the apex, cuneate to rounded at the base and often asymmetric, glabrous; lateral nerves in 8–13 pairs, prominent and often drying yellowish beneath; venation prominently raised-reticulate. Inflorescences 9–30(–60) cm. long, with spreading or ascending main branches up to 35 cm. long, ± glabrous or ferruginous (in life and dry) pubescent, sometimes only at the nodes; flowers in dense clusters on very short side branchlets, pedicels said to be 'brown at base but white above joint' 0–3 mm. long, lengthening to 5 mm. in fruit. Sepals drying cream or brown, 4.5 mm. long, 2.8 mm. wide, the outer strongly overlapping the inner, petaloid, glabrous to densely dark ferruginous-pubescent outside, but always ciliate; petals white or cream, sometimes tinged pink, obovate, 4–5 mm. long, 2.5 mm. wide, shaggy ciliate below, scale 3 mm. long. Stamens 14–18, with filaments white hairy. Fruit ± globose, 1.2–1.5 cm. long, 1.2–1.8 cm. wide, usually only one mericarp developing, the two aborted ones persistent at the base, glabrous. Seed 1 cm. long.

var. **kilimandscharica**

Calyx glabrous (save for marginal ciliae) or slightly puberulous outside. Fig. 22.

UGANDA. Mbale District: Bufumbo, Busano track, Nkokonjero, Oct. 1942, *Dale* U348 & Elgon, 26 Oct. 1923, *Snowden* 796!; Mengo District: Mulange, Oct. 1922, *Dummer* 5582!

KENYA. Nandi District: Kaimosi, Sirwa Farm, near Yala R. bridge on Kapsabet–Kisumu road, Nov. 1971, *Tweedie* 4158!; Kiambu District: Muguga Forest, 15 Feb. 1967, *Kirrika* 493!; N. Kavirondo District: Kakamega Forest, Kakamega–Kaimosi road, near forester's house, 15 Oct. 1953, *Drummond & Hemsley* 4765!

TANZANIA. Mbulu District: Tlambi, 21 Sept. 1966, *Carmichael* 1234!; Lushoto District: above track past Amani lake, 15 Apr. 1968, *Renvoize & Abdallah* 1481!; Morogoro District: Nguru Mts., Turiani, NW. slope of Mkobwe, 29 Mar. 1953, *Drummond & Hemsley* 1864!

DISTR. **U** 3, 4; **K** 3–5, 6 (Ngong); **T** 2, 3, 6; E. Zaire and Ethiopia

HAB. Evergreen forest, including gorge forest and downward extensions of upland rain-forest; 1100–2250(–?2400) m.

FIG. 22. *DEINBOLLIA KILIMANDSCHARICA* var. *KILIMANDSCHARICA* — **1**, female flowering branchlet, × $^{2}/_{3}$; **2**, male flower with sepal and petals removed, × 6; **3**, petal from male flower, × 6; **4**, female flower with sepal and petals removed, × 6; **5**, petal from female flower, × 6; **6**, vertical section of female flower, × 8; **7**, ovary from female flower, × 16; **8**, fruiting branchlet, × $^{2}/_{3}$. 8, from *Drummond & Hemsley* 2624. Drawn by Maureen Church.

var. **adusta** (*Radlk.*) *Verdc.* var. et comb. nov. Type: Tanzania, Lushoto District, W. Usambara Mts., Kwai, *Albers* 107 (B†, holo.)

Calyx sparsely to densely dark ferruginous pubescent.

TANZANIA. Lushoto District: W. Usambara Mts., Gologolo, 26 Oct. 1961, *Mgaza* 450!; Morogoro District: Uluguru Mts., 21 Nov. 1932, *Wallace* 471!; Iringa District: Tapu, Nov. 1953, *Carmichael* 281!

DISTR. **T** 3, 6, 7; N. Malawi

HAB. Evergreen forest, *Podocarpus-Ocotea* etc.; 900–1950 m. (see note)

SYN. *D. adusta* Radlk. in E.P. IV, 165: 681 (1932); T.T.C.L.: 556 (1949)

NOTE. I have only kept this up since the tendency to have a hairy calyx has a geographical trend but many specimens in these eastern arc forests have glabrous calyx-lobes; whether it is a subspecies in the making or two taxa previously separate now introgressing is not clear. Davies did not recognise it and gives the lower altitudinal level of the species as 600 m.

3. **D. borbonica** *Scheff.*, Tijdschr. Nederl. Ind. 31 (Obs. Phytog.): 17 (1869) & in Flora 52: 306 (1869); Radlk. in E.P. IV, 165: 674 (1932); Exell in F.Z. 2: 522 (1966); Capuron in Mém. Mus. Nat. Hist. Nat., Paris, 19: 72 (1969). Type: Indonesia, Java, from a tree grown at Bogor (no. 14528) from material said to have been sent from Réunion (BO!, lecto.)*

Shrub or small tree 0.6–4.5 m. tall with simple stem, monoecious; branchlets brownish or blackish, golden hairy when young, later glabrescent. Leaves terminal with petiole 7–15 cm. long; rhachis 20–42 cm. long; petiolules 0–4 mm. long; leaflets (3–)6–10-jugate, alternate or subopposite, oblong-elliptic to elliptic or sometimes obovate to oblanceolate, 5–23 cm. long, 1.9–8.5 cm. wide, mostly rounded or subacute at the apex, cuneate to rounded at the base or rarely subcordate, often densely ferruginous pubescent beneath when very young, but mostly glabrous save for a few adpressed hairs on the main nerves beneath but rarely ± densely ferruginous pubescent beneath; lateral nerves in 5–9(–12) pairs, venation very prominent on both sides, particularly beneath. Inflorescences terminal, 30–32(–60) cm. long, shortly branched at the base or much branched; ultimate cymules (2–)4–6-flowered; rhachis velvety ferruginous; pedicels 1–3 mm. long. Calyx densely dark brown or silvery to creamy-ferruginous pubescent outside; sepals elliptic to round, 4–5 mm. long, 2.5–4.5 mm. wide; petals white, drying red-brown, elliptic, 5.5–6 mm. long, 2.8–3 mm. wide, with long white cilia, scale 4 mm. long. Stamens 15–17; filaments white, hairy; anthers pale yellow; staminodes short. Ovary densely pubescent or glabrous in a Madagascan form (see general note); style and stigma pale green; ovary rudiment very reduced. Fruiting inflorescence up to 40 cm. long; mericarps 2–3, yellow or white, red or golden brown, ovoid or subglobose, 1–1.4 cm. long, densely pale brown hairy when young, becoming thinly hairy; flesh white, mucilaginous.

forma **glabrata** *Radlk.* in Sitz. Akad. Wiss. München 8: 369 (1878) & in P.O.A. C: 250 (1895) & in E.P. IV, 165: 675 (1932); T.T.C.L.: 556 (1949); K.T.S.: 509 (1961); Haerdi in Acta Trop., suppl. 8: 124 (1964); Capuron in Mém. Mus. Nat. Hist. Nat., Paris, 19: 74 (1969). Types: Comoro Is., Mayotte, *Boivin* 3358 (P, syn.) & Zanzibar, *Boivin* (P, syn.) & *Hildebrandt* 1159 (B, syn., W, isosyn.)

Leaflets ± glabrous or with sparse hairs on nerves beneath. Calyx pale brown pubescent.

* "Crescit in insula Bourbon unde a b. Diard in hortum communicata". As Capuron points out *D. borbonica* is not known from Mauritius or Réunion so the origin of the type is in doubt. I have chosen as lectotype a sheet at Bogor bearing the number 14528 and annotated by Radlkofer in Jan. 1903. I am grateful to R. Johns for arranging for this sheet to be sent to Kew on loan.

KENYA. Machakos/Kitui District: Tsavo National Park East, 5 km. between Tsavo Safari Camp and Athi R., 25 Dec. 1987, *S.A. Robertson* 5024!; Teita District: Voi–Mwatate road, bridge over the Voi R., 12 Sept. 1953, *Drummond & Hemsley* 4277!; Kilifi District: Arabuko-Sokoke Forest, Jilori, 26 Nov. 1961, *Polhill & Paulo* 861!

TANZANIA. Pangani District: Jasini Forest, 17 Oct. 1955, *Tanner* 2304!; E. Mpwapwa, 24 Aug. 1930, *Greenway* 2460!; Iringa District: near Great Ruaha R., 11 km. W. of Kidatu Bridge, 16 July 1970, *Thulin & Mhoro* 449!; Zanzibar, Mazizini [Massazine], 10 Sept. 1959, *Faulkner* 2358!

DISTR. **K** 4, 7; **T** 2, 3, 5–8; **Z**; **P**; S. Somalia, Malawi, Mozambique and Comoro Is.

HAB. Riverine *Acacia* thorn bush and evergreen thicket, *Combretum-Acacia* woodland, low evergreen forest on limestone outcrops; 0–1050 m.*

SYN. *D. nyikensis* Baker in K.B. 1897: 249 (1897); Radlk. in E.P. IV, 165: 677 (1932); Exell in F.Z. 2: 522 (1966). Type: Malawi, Nyika Plateau, between Kondowe and Karonga, *Whyte* (K!, holo.)

[*D. borbonica* sensu Exell in F.Z. 2: 522 (1966); Vollesen in Opera Bot. 59: 58 (1980); Beentje, K.T.S.L.: 416, map (1994), *non* Scheff. sensu stricto]

forma **subcordata** *Verdc.*, forma nova a formis ceteris foliolis basi subcordatis, usque 20 × 10.5 cm. subtus molliter ferrugineo-pubescentibus differt. Typus: Tanzania, Lushoto District, Korogwe, Mombo Forest Reserve, *Ngonyani* 49 (K!, holo., EA, iso.)

Leaflets oblong, up to 20 cm. long, 10.5 cm. wide, rounded at the apex, subcordate at the base, softly ferruginous pubescent beneath and with some sparse hairs above particularly on the midrib. Inflorescences with pale brown pubescence as in forma *glabrata*.

TANZANIA. Lushoto District: Korogwe, Mombo Forest Reserve, site 6, 13 Oct. 1971, *Ngonyani* 49!

DISTR. **T** 3; not known elsewhere (see note)

HAB. ?Evergreen forest; ± 400 m.

GENERAL NOTE. Radlkofer was, I think, justified in establishing a forma *glabrata* for the East African plant, particularly as the origin of the type plant is not known – it seems impossible it could have come from Réunion. Capuron has two other forms from Madagascar, one with a glabrous ovary. *Bally* 1977 (Kenya, Tana River District, Garissa) has been referred to var. *minor* Radlk. (E.P. IV, 165: 675 (1932); type: Somalia, Baraawe [Barana], *Hildebrandt* 1325 (B†, holo.), described as 'foliola minora, rigide chartacea, insigniter reticulato-venosa' and littoral in habitat); its status is not known.

Capuron has sunk *Omalocarpus macrophyllus* Choux (type: Madagascar, Boina, Ankarafantsika, *Perrier de la Bâthie* 13222 (P, holo., K!, iso.)) into *D. borbonica* as forma *pilosula* but it is much more like *D. kilimandscharica* and the presence of *D. borbonica* in Madagascar is, I think, doubtful.

22. SAPINDUS

L., Sp. Pl.: 367 (1753) & Gen. Pl., ed. 5: 171 (1754); Radlk. in E.P. IV, 165: 630 (1932)

Trees. Leaves usually rather pale green, pinnate; leaflets in 2–8 pairs, entire; rhachis with raised lines or sometimes winged. Flowers in terminal and axillary panicles containing both male and female flowers but one sex often predominating. Flowers regular; sepals 5, 2 short and 3 larger, concave, imbricate; petals 5, with or without scales, clawed and hairy at base; disk complete, cupuliform, undulate and shallowly lobed, glabrous. Stamens 8; filaments stout, with lower half hairy; anthers oblong-ovate, auriculate, subsessile in bud; staminodes similar but smaller. Ovary 3-lobed, glabrous or hairy; style entire; ovary rudiment present in male, 3-lobed. Fruit a 1–2(–3)-lobed drupe, the abortive mericarp(s) remaining at base.

A genus of ± 14 species in Asia and America; 2 cultivated in Africa.

* *Johnston* from Kilimanjaro has 5000' on the label (1500 m.) but this is a generalised label and I think it unlikely it occurs so high. Davies gives 2000 m. possibly from some record outside the Flora area.

Some explanation is needed about the gender of the genus. Since Linnaeus has both '*saponaria*' and '*trifoliata*' it might be thought to be feminine but most* subsequent authors have used '*trifoliatus*'. Linnaeus spelt *saponaria* with a capital S showing it was derived from the generic name *saponaria* and thus a noun in apposition not an adjective; the spelling *trifoliata* is an error and although repeated by Linnaeus in Sp. Pl., ed. 2 he has there a third species *spinosus* which shows he meant *Sapindus* to be masculine.

Leaflets in 2–6 pairs, elliptic-lanceolate to lanceolate, broadest in the lower third(–half), 7–17.5 cm. long, 2.5–6 cm. wide; sepals glabrous save at the base; young ovary glabrous, black when dry . . . . 1. *S. saponaria*

Leaflets in (1–)3(–4) pairs, elliptic-oblong to obovate, broadest at ± the middle, (5–)10–20 cm. long, (3–)5.5–11 cm. wide; sepals silky-hairy; young ovary densely pubescent . . . . 2. *S. trifoliatus*

1. **S. saponaria** *L.*, Sp. Pl.: 367 (1753); Radlk. in E. & P. Pf. III, 5: 316, fig. 164 (1895) & in E.P. IV, 165: 639, fig. 14 (1932); Burkill, Dict. Econ. Prod. Malay Pen.: 1959 (1935); T.T.C.L.: 561 (1949); U.O.P.Z.: 441, fig. on 442 (1949); Dale, Introd. Trees Uganda: 64 (1953); C.F.A. 2: 79 (1954); Jex-Blake, Gard. E. Afr., ed. 4: 125 (1957); Hauman in F.C.B. 9: 332 (1960); Fouilloy & Hallé in Fl. Cameroun 16: 58, t. 12/1–11 (1973) & in Fl. Gabon 23: 58 t. 12/1–11 (1973); Leenh. in Fl. Males., ser. 1, 11: 714, fig. 70 (1994). Lectotype, chosen by Pennington in Reg. Veg. 127: 84 (1993): Plukenet, Phytographia, t. 217/7 (1692) (lecto.)

Tree 8–18 m. tall with rounded crown; bark ochraceous, smooth on branches but very rough and corrugated on trunk, scaling off in large flakes to reveal brown patches; slash pale orange; twigs grey, glabrous, with deep brown lenticels. Leaves 20–30(–60) cm. long; petiole (1–)4–8 cm. long, minutely pubescent; rhachis pale green or whitish, 14–40 cm. long, winged or with raised lines; petiolules (1–)3–5 mm. long; leaflets in 2–6 pairs, subopposite, elliptic-lanceolate to lanceolate, 7–17.5 cm. long, 2.5–6 cm. wide, often ± asymmetrically curved, narrowed to an acute or acuminate apex, cuneate at the base, glabrous above apart from the midrib, shortly pubescent beneath and with sessile glands; lateral nerves in 9–12 pairs. Inflorescences complex panicles of cymules, 30–75 cm. long, 20 cm. wide; axes puberulous; pedicels ± absent or 1–2 mm. long. Flowers creamy white or greenish white, small, slightly aromatic; sepals 2–3 mm. long; petals rhomboid, 1.5 mm. long, with long hairs; disk glabrous. Fruit brown drying blue-black, with 1(–3) globose lobes, 1.7–2 cm. diameter, gelatinous when fresh, glabrous; pedicels 5–10 mm. long. Seeds black, globose, 1.2 cm. diameter. Fig. 23/1–3.

UGANDA. Mengo District: Entebbe Botanic Gardens *fide* Dale

TANZANIA. Mwanza District: Bukumbi [Buhumbi], Kashishi, 12 Feb. 1953, *Tanner* 1214!; Lushoto District: Mombo, 1 Aug. 1960, *Semsei* 3064!; Uzaramo District: Dar es Salaam, 20 July 1969, *B.J. Harris* 2978!

DISTR. **U** 4; **T** 1, 3–6; native of Central and South America widely cultivated throughout the tropics

HAB. No evidence that it has become naturalised but so widely planted it has been included; used as an avenue tree, also seen on open land, grassland on deep sand, etc.; ± 0–1150 m.

NOTE. The soap-berry tree produces large quantities of saponin in the epicarp of the fruit and also in the roots; the berries are much used as soap. The timber is hard and close-grained but not durable. The seeds are widely used for the game mbao and for necklaces.

* Nicolson et al. in Reg. Veg. 119: 237 (1988) treat it as feminine, I think wrongly. I have made a proposal to conserve the gender of *Sapindus* as masculine (Taxon 46: 360 (1997)).

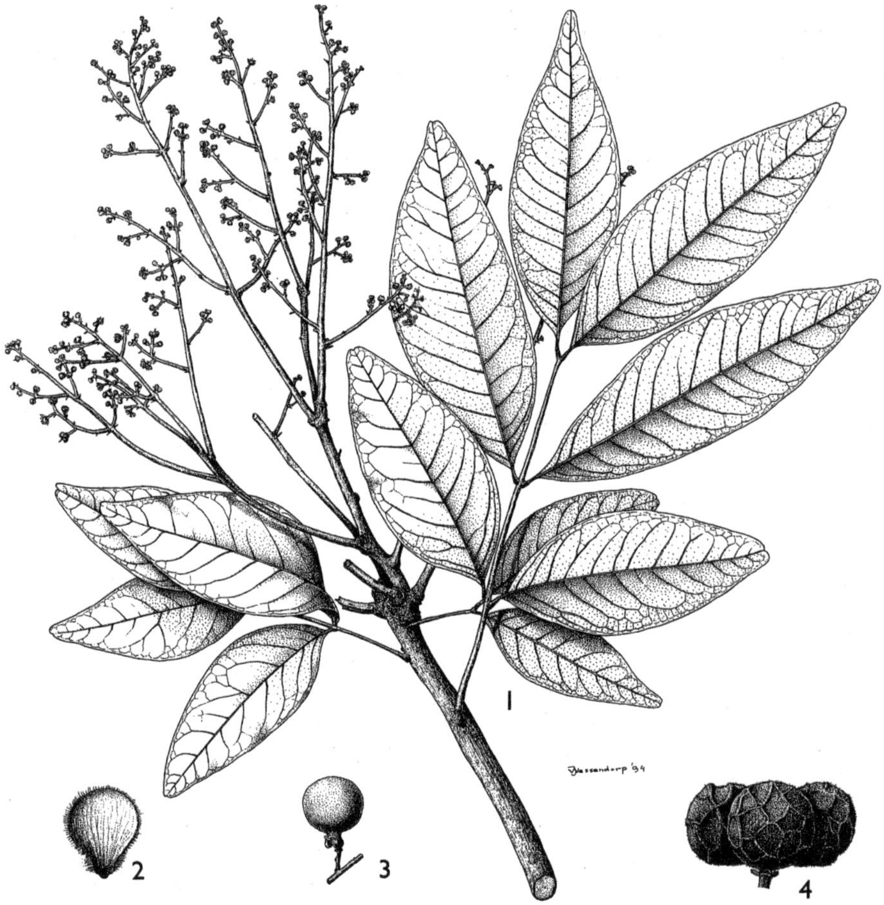

FIG. 23. *SAPINDUS SAPONARIA* — **1**, flowering branch, × $^1/_2$; **2**, petal, × 6; **3**, fruit, × 1. *SAPINDUS TRIFOLIATUS* — **4**, fruit, × 1. 1, 2, from *Iboet* 255; 3, from *Schodde* 2779; 4, from *Kostermans* 25836. Reproduced with permission from Fl. Males., ser. 1, 11, fig. 70 (1994).

2. **S. trifoliatus** *L.*, Sp. Pl.: 367 (1753), as '*trifoliata*'; Radlk. in E.P. IV, 165: 656 (1932); Burkill, Dict. Econ. Prod. Malay Pen.: 1959 (1935); T.T.C.L.: 561 (1949); Dale, Introd. Trees Uganda: 64 (1953). Type: Rheede, Hort. Malab. 4: 43–44, t. 19 (1683) (holo.)

Tree to 25 m.; branchlets purplish brown, lenticellate. Leaves 15–25(–35) cm. long; leaflets subopposite, in (1–)3(–4) pairs, elliptic-oblong to elliptic-lanceolate or obovate, (5–)10–20 cm. long, (3–)5.5–11 cm. wide, emarginate to ± acuminate at the apex, acute to truncate or rounded at the base, glabrous save for some pubescence on midrib above and beneath; lateral nerves in 8–15 pairs. Panicles 6–15 cm. long, with spreading branches, velvety brown-pubescent; sepals elliptic, 4 mm. long, silky tomentellous outside; petals narrowly rhomboid, 4 mm. long, the dorsal surface densely long-white-pubescent. Ovary pubescent; disk hairy. Fruit brown, 1.6–2.2 cm. wide, with (1–)3 subglobose hairy lobes. Fig. 23/4.

UGANDA. Busoga District: Vunamba Nile Forest, Aug. 1945, *Wakabi* 22!; Mbale Station, July 1945, *Dale* U418!
KENYA. Kilifi District: Rabai, *van Someren* 316
TANZANIA. Bagamoyo boma, 22 July 1960, *Willan* 520!; Uzaramo District: Dar es Salaam Botanic Garden, 21 Jan. 1977, *Ruffo* 1013!; Lindi District: Nachingwea, *Berrie* H191/64
DIST. **U** 3, **K** 7, **T** 6, 8; native of India now widely cultivated
HAB. Probably not naturalised anywhere; ± 0–1200 m.

NOTE. The soap-nut tree; the fruits of this are also used for soap.

23. ALLOPHYLUS

L., Sp. Pl.: 348 (1753) & Gen. Pl., ed. 5: 164 (1754); Radlk. in E.P. IV, 165: 455 (1932); Leenh. in Blumea 15: 301–358 (1967) & in Fl. Males., ser. 1, 11: 459–465 (1994)

Schmidelia L., Syst. Nat., ed. 12, 2: 274 (1767) & Mant. Pl.: 10, 67 (1767), *nom. illegit.*, *non* Boehm. (1760)

Trees, shrubs or woody climbers, monoecious or dioecious. Leaves 1–3(–4)-foliolate, often minutely glandular beneath. Flowers small, slightly irregular, sessile or shortly pedicellate, in several-flowered cymules which are densely or more openly placed along simple raceme-like axes or branched axes thus appearing paniculate; cymules sessile or shortly stalked. Sepals 4, unequal, imbricate; petals 4, spathulate with internal ciliate scale; disk unilateral, 2–4-lobed. Stamens 8, placed to one side of the disk; reduced to staminodes in female flowers. Ovary 2–3-locular, deeply 2–3-lobed, the lobes subglobose, with 1 basal ovule in each; style semigynobasic, 2–3(–4)-fid. Fruit usually red, 1–2(–3)-lobed and seeded, drupe-like, the abortive lobes not persisting. Seeds straw-coloured, ellipsoid to globose, without an aril.

There is much disagreement about how many species should be recognised in this genus which is of exceptional difficulty; nothing is known of the reproductive biology so it is not possible to suggest a reason for the poor segregation of the taxa. Once the reason is discovered the course of action to be followed may become clearer. Suggestions that the difficulties are due to frequent hybridisation are unsubstantiated and annotation labels giving precise hybrid identifications are naive in the extreme. Usually 150–200 species are recognised, widely distributed in the tropics and subtropics of both Old and New Worlds but Leenhouts took a radical step and considered there to be only one species *A. cobbe* (L.) Räusch. and suggested that in each area races could be recognised without formal taxonomic status. Mabberley (The Plant Book, ed. 2: 25 (1997)) accepts that there is one very polymorphic species "which at a local level can be subdivided into c. 175 apparently distinct species". Capuron has also followed this in his account of the Madagascan Sapindaceae (Mém. Mus. Nat. Hist. Nat., Paris, 19: 53–68 (1969)) using the name *A. cobbe* followed by the usually accepted species name in brackets to produce an informal system of dealing with the entities. To many botanists, however, Leenhouts's suggestion is quite unacceptable. Exell & Sousa (in Fl. Moçamb. 51: 3–16 (1973)), Vollesen (Fl. Ethiopia 3: 498–502 (1990)), Beentje (K.T.S.L.: 412–413 (1994)) and Pennington (as yet unpublished account of the South American species) have ignored it. One can understand Leenhouts's reasons. I found taxa of *Allophylus* in Sri Lanka and Taiwan which seem virtually indistinguishable from bushland species in East Africa and when shown to someone dealing with South American species were thought to be a Bolivian taxon. Clearly this is not easy to accept. On the other hand, anyone shown the unifoliolate coastal shrub *A. pervillei* and the trifoliolate upland forest tree *A. abyssinicus* is unlikely to accept them as the same species and biologically I am sure they are not. I have accepted 18 species in East Africa. Certainly it would be infinitely easier to follow Leenhouts but it does not seem very useful. Undoubtedly the forest species are easier to delimit than the dry country and bushland species. I have resisted treating any of the East African bushland taxa as variants of *A. cobbe* sensu stricto even if ± identical since this would imply that they were closer to this Sri Lanka species than to the East African forest taxa.

Joseph Njung'e (Chemical and pollen study of genus *Allophylus* (Sapindaceae) M.Sc. Dissertation, Reading University) examined 31 specimens representing only 12 species (the names on the herbarium sheet labels being accepted at face value). Information from the chemical data was insufficient to help with specific delimitation. Pollen grains were found to be homogenous in type and aperture number but the exine sculpture was found to be variable and falls into 7 types varying from poorly developed reticulate to retirugulate thence to well-developed rugulate or rugulate-striate. This is somewhat unexpected and needs expanding using some hundreds of specimens.

Several collectors mention scented flowers and their attraction for flies; most of the species could be fly-pollinated.

It is not possible to produce a key to the species of *Allophylus* which will work with every specimen and the following is but a guide. Inflorescences unbranched is used in a special sense in this genus by all workers; the cymules themselves are of course short branches of the main rhachis but these are ignored – unbranched indicates that the main rhachis of cymules is itself unbranched – branched indicates that there are 2–many branches each bearing numerous cymules.

1. Leaves 1-foliolate 2
   Leaves 3-foliolate but sometimes lateral leaflets very small 5
2. Branchlets ferruginous-velvety-tomentose; leaflets velvety tomentose or pilose (**T** 4, 6–8) .... 14. *A. congolanus*
   Branchlets much less hairy; leaflets glabrous save for hairy domatia beneath in some leaves 3
3. Cymules shortly stalked the stalks 2–4 mm. long; hairy domatia absent (**T** 3, E. Usambara Mts.) .... 16. *A. grotei*
   Cymules sessile or subsessile with stalks to 1.2 mm. long 4
4. Shrub or occasionally a small tree 1–4(–5) m. tall, mostly in coastal thicket; inflorescences unbranched; petiole 0.5–1.5 cm. long; leaves ± thick .... 18. *A. pervillei*
   Tree 5–12 m. tall of rain-forest; inflorescences sometimes ± branched; petiole 1.5–3 cm. long; leaves ± thin .... 17. *A. melliodorus*
5. Lateral leaflets very small in proportion to terminal leaflet, not exceeding 4 cm. long 6
   Lateral leaflets much larger in proportion to terminal leaflet, but all leaflets may be small and well under 4 cm. in some species (e.g. *A. rubifolius*) 8
6. Leaflets glabrous save for hairy domatia beneath; lateral leaflets 1.5–2 × 0.8–1.3 cm. .... 18. *A. pervillei*
   Leaflets densely hairy 7
7. Terminal leaflet 20–30 × 9–15 cm., tapering to a narrow acute apex, densely velvety pilose; lateral leaflets 1.5–4 × 0.5–2 cm.; petioles, inflorescence-rhachis and fruits with long spreading hairs; cymules regularly closely placed ± 0–2 mm. apart; inflorescences ± 6 cm. long (**T** 6, **P**) .... 15. *A. vestitus*
   Terminal leaflet 5–17.5 × 3–10.5 cm., mostly acute at the apex but not so strongly tapering, densely hairy at least beneath but hairs shorter and less dense above; lateral leaflets 1–3 × 1–3 cm.; petioles and inflorescence-rhachis more shortly pubescent and fruits with sparse ± adpressed hairs; cymules more irregularly spaced 1–6 mm. apart; inflorescences 6–20 cm. long (**T** 4, 6–8) .... 14. *A. congolanus*
8. Inflorescences branched or unbranched, the rhachis mostly densely hairy with short and long hairs, the longer over 1 mm. long; leaves drying dark purplish brown above, dark green beneath, the midrib above and beneath mostly with similar hairs to inflorescence-rhachis; pedicels glabrous at least in upper part (**U** 1, 2, 4, **T** 1; 1200–2250 m.) .... 6. *A. pseudopaniculatus*
   Without above characters combined 9

9. Usually a medium to fairly large tree of upland evergreen forest and derivatives at 1000–2550 m.; indumentum of branchlets, leaves, petioles, inflorescences, etc. a fine ferruginous tomentum, with hairs under 0.1 mm. long; inflorescences 5–22 cm. long, with 2–10 branches; leaflets 5.5–21 × 3–11 cm., drying dark brown above, paler brown beneath (widespread) · · · · · 1. *A. abyssinicus*
If a forest tree then indumentum not of such short hairs, or absent · · · · · 10
10. Inflorescences branched · · · · · 11
Inflorescences predominantly unbranched or few with 1–2 short branches · · · · · 17
11. Inflorescences very well developed and much-branched, 6–28 cm. long; leaflets large, drying pale bright green, 7.5–25 × 3.5–10.5 cm. (some leaflets only 4 × 1.3 cm. in *A. delicatulus*) · · · · · 12
Without these two characters combined · · · · · 15
12. Inflorescences very graceful; cymules 1–2-flowered with slender pedicels up to 2.5 mm. long; plant virtually glabrous; leaflets up to 11 × 4 cm., narrowly acuminate (**T** 6, Mahenge) · · · · · 7. *A. delicatulus*
Inflorescences coarser and without other characters combined · · · · · 13
13. Inflorescence-branches 2.5–4.5 cm. long, held at right angles to main rhachis (**U** 2, Toro, Ruwenzori) · · · 10. *A. sp.*
Inflorescence-branches if at right angles then longer · · · · · 14
14. Cymule-rhachises nodular with pedicel bases very short and closely placed; indumentum sparse, white (**K** 7, **T** 3, 8) · · · · · 2. *A. chirindensis*
Cymule-rhachises with pedicel bases longer and diverging; indumentum ferruginous (**U** 1–3) · · · · 3. *A. dummeri*
15. Leaflets glabrous save for hairy domatia beneath, drying dark brown and often glossy above, paler brown or greenish beneath, subcoriaceous, subentire or crenae and teeth shallow; inflorescence-rhachises usually rather short, densely shortly grey or usually ferruginous pubescent (widespread) · · · · · 4. *A. africanus* var. *africanus**
Leaflets ± hairy · · · · · 16
16. Leaflets predominantly rounded, truncate or obtuse (rarely acute) at the apex; cymules of inflorescence very densely placed so that inflorescence is a solid continuous mass of flowers save at extreme base; venation closely reticulate and ± prominent (**T** 1, 2, 4, 5, 7) · · · · · 5. *A. fulvotomentosus*
Leaflets mostly acute and inflorescences not so dense · · · · · 4. *A. africanus* var. *griseotomentosus***

* Some specimens of *A. tanzaniensis* with 1 or 2 short side branches will key here but the leaves are mostly pale green on drying and inflorescence-rhachis glabrous or only puberulous (**T** 5–8)

** Branched forms of some species with normally unbranched inflorescences will also key here; see note at end.

17. Leaflets drying leaden or dark brown above with venation pale, with strongly reticulate prominent venation above and beneath, rather thick, 4–15.5 × 2–6 cm., glabrous or ferruginous hairy; flowers often rather larger than is usual in the genus, 3–3.5 mm. wide (**T** 4, 7; 1500–2250 m.) ············ 8. *A. chaunostachys*
   Not as above; leaflets less prominently reticulate ···················· 18
18. Leaflets mostly large, ovate-lanceolate or elliptic-lanceolate, (2.5–)6–22 × (0.9–)2.5–11 cm., with laterals mostly distinctly asymmetric, ± glabrous or distinctly ferruginous hairy (mostly drying pale or bright green on both surfaces) ·············· 9. *A. ferrugineus*
   Leaflets smaller, under 10 cm. long; indumentum mostly white or grey ···················· 19
19. Leaflets practically glabrous ···················· 20
   Leaflets pubescent to velvety ···················· 21
20. Leaflets elliptic or oblong, entire or coarsely shallowly crenate; calyx glabrous; inflorescence-rhachis glabrous or minutely puberulent, rarely pubescent (**T** 5–8; 350–1400 m.) ···················· 11. *A. tanzaniensis*
   Leaflets obovate; inflorescence-rhachis often pubescent, also petioles etc. (widespread throughout East Africa, but particularly in coastal areas) ····· 12. *A. rubifolius* var. *alnifolius*
21. Leaflets up to 10 × 4.5 cm., thin, drying purplish brown above, bright green beneath with adpressed ± bristly pubescence on venation beneath and patterned with small tufts on midrib and along lateral nerves in most vein-axils; fruits with very scattered hairs (**T** 7 – a specimen from Iringa, amongst granite blocks) ··················· 13. *A. richardsiae*
   Not exactly as above; if with patterns of hair-tufts beneath and from same area then leaflets smaller and obovate (very widespread and commonspecies) ···· 12. *A. rubifolius**

NOTE. There are numerous intermediates between the various varieties of *A. rubifolius* (see key on p. 89) and *A. ferrugineus* and some forms of *A. africanus*. The velvety variety *A. rubifolius* var. *dasystachys* frequently has branched inflorescences ('*A. stachyanthus*') and merges with *A. africanus* var. *griseotomentosus*. These can only be named by comparison in the herbarium and judicious use of geographical data in conjunction with the knowledge that leaf-size and indumentum and inflorescence-branching are extremely variable.

1. **A. abyssinicus** (*Hochst.*) *Radlk.* in E. & P. Pf. III, 5: 313 (1895) & in Sitz. Akad. Wiss. München 38: 223 (1909); Z.A.E.: 476 (1912); Baker f. in J.B. 57: 186 (1919); Radlk. in E.P. IV, 165: 534 (1932); F.P.N.A. 1: 517 (1948); T.T.C.L.: 554 (1949); I.T.U., ed. 2: 375 (1952); Wimbush, Cat. Kenya Timbers, ed. 2: 28 (1957); Hauman in F.C.B. 9: 303 (1960); K.T.S.: 504 (1961); F.F.N.R.: 223 (1962); Exell in F.Z. 2: 498 (1966); Hamilton, Field Guide Uganda For. Trees: 97 (1981); Troupin, Fl. Rwanda 2: 298 (1983); Vollesen in Fl. Ethiopia 3: 499, fig. 125.7 (1990); Beentje, K.T.S.L.: 412, map (1994). Type: Ethiopia, Semien, Addi Dschoa, *Schimper* I. 377 (P, holo., BM!, C!, K!, iso.)

* Here also will key simple inflorescence forms of *A. africanus* var. *griseotomentosus*.

Medium to fairly large tree, rarely shrubby, (4.5–)6–21(–25) m. tall and with slender bole 30–90 cm. diameter, sometimes slightly buttressed at the base (e.g. in Chyulu Hills); crown dark green, open to dense, irregular or rounded; bark grey or greyish white, smooth or slightly rough or granular, sometimes peeling in small to large flakes; slash pale brownish pink or pale orange darkening to reddish brown and smelling of pepper, also described (*Ede* 1, **T** 7) as off-white with pinky brown green-edged lines; sapwood white; young branches minutely ferruginous tomentellous and densely lenticellate. Leaves trifoliolate, drying dull and dark above, thinly coriaceous; petiole 2.5–10.5 mm. long, minutely tomentellous or less often more distinctly pubescent; leaflets elliptic to obovate, (2.8–)5.5–21 cm. long, (1.2–)3–11 cm. wide, acute to acuminate at the apex, attenuate to cuneate at the base, shallowly crenulate to toothed, adpressed velvety pilose when very young, later glabrescent or glabrous to minutely puberulous, less often pubescent; tufts of hairs in axils of nerves and veins usually conspicuous beneath. Inflorescences 5–22 cm. long, with 2–10 branches, some again branched, usually finely tomentose; cymules subsessile; pedicels 1–3 mm. long. Flowers said to be lily-scented or scentless; outer sepals elliptic, 1.2–1.5 mm. long, 1 mm. wide, puberulous, inner elliptic to round, 1.5–2 mm. long, 1.5 mm. wide; petals dirty white or yellowish, 1.5 mm. long, 1 mm. wide. Filaments 2–3 mm. long, glabrous. Style 2–3 mm. long, ± glabrous; stigma 1–1.5 mm. long. Fruit red or orange, subglobose, oblong or elliptic, 6–8 mm. long and wide, with very sparse hairs. Seed ellipsoid, 5–5.5 mm. long, 5.5–6 mm. wide.

UGANDA. Mt. Ruwenzori, Dec. 1938, *Sangster* 486!; Mbale District: Sebei, N. Elgon Forest Reserve, Kere, 10 Jan. 1963, *Styles* 298!; Mubende District: Kasambya, Dec. 1937, *Mr. & Mrs. Sitwell* in *Eggeling* 3470!

KENYA. Northern Frontier Province: Mt. Kulal, 9 June 1960, *Oteke* 96!; Trans-Nzoia District: NE. Elgon, Kisano R., Nov. 1961, *Tweedie* 2243!; Machakos District: Ol Doinyo Sapuk [Ol Donyo Sabuk], 29 Apr. 1980, *M.G. Gilbert* 5941!

TANZANIA. Arusha District: S. Meru, eastern sawmills, 22 Mar. 1951, *McCoy-Hill* 6!; Lushoto District: Mkusu Forest Reserve, 11 Sept. 1969, *Shabani* 458!; Kilosa District: Ukaguru Mts., near Mandege Forest Station nursery, 9 Aug. 1972, *Mabberley* 1434!

DISTR. **U** 1–4; **K** 1, 3–5, ?6; **T** 2, 3, 6 7; E. Zaire, Rwanda, S. Sudan, Ethiopia, Malawi, Mozambique and Zimbabwe

HAB. Rain-forest, montane forest edges, evergreen thicket, woodland, sometimes riparian; (650–)1050–2550(–?2850) m.

SYN. *Schmidelia abyssinica* Hochst. in Flora 26: 80 (1843)
[*S. africana* sensu Baker in F.T.A. 1: 421 (1868) pro parte, *non* (P.Beauv.) DC.]

NOTE. This is one of the more distinctive species. The timber is almost valueless due to lack of durability but has been used for cheap joinery and boxes and doubtless more extensively for local purposes. Exell states that the stigma is 3-fid but in the Flora area it is nearly always 2-fid.

2. **A. chirindensis** *Baker f.* in J.L.S. 40: 48 (1911) & in J.B. 57: 188 (1919); Exell in F.Z. 2: 505 (1966); Exell & Sousa in Fl. Moçamb. 51: 12 (1973). Type: Zimbabwe, Chirinda Forest, *Swynnerton* 112 (BM, holo., K! (2 sheets), iso.)

Much-branched shrub or tree 3–15 m. tall; bark silvery grey; branchlets pale greyish, lenticellate, glabrous. Leaves trifoliate; petiole 3–16 cm. long, grooved, glabrous or minutely puberulous; petiolule of terminal leaflet 1.2–2 cm. long; leaflets drying pale bright green, elliptic or elliptic-oblong, 9–19.5 cm. long, 3.5–9 cm. wide, rounded to acuminate at the apex, narrowly cuneate at the base, subentire, coarsely undulate-crenate and shortly denticulate (the teeth hydathodes) or serrate, thin, glabrous or minutely puberulent on nerves beneath with white hairs and sometimes with tufts in the nerve axils beneath; lateral nerves in 7–13 pairs. Inflorescences up to 28 cm. long, much branched, the branches 10–14 cm. long, glabrous or sparsely white-puberulous; peduncle 14–20 cm. long; cymules few-flowered, sessile; pedicels 3–4 mm. long, glabrous. Flowers white or yellow; outer sepals round, 1.8 mm.

diameter, inner elliptic, 1.8 mm. long, 1 mm. wide, glabrous; petals spathulate, 1.5 mm. long, 1 mm. wide, with hairy scale. Filaments 2 mm. long; staminodes in female flowers 0.6 mm. long. Ovary 2-lobed, minutely pubescent; style 1.5 mm. long. Fruit red, subglobose, 7 mm. in diameter, minutely puberulous when young or ± glabrous.

KENYA. Kwale District: Shimba Hills, Lango ya Mwagandi [Longo Mwagandi] Forest, 20 May 1968, *Magogo & Glover* 1098!

TANZANIA. Lushoto District: E. Usambara Mts., Kisiwani, 2 May 1939, *Greenway* 5823! & Amani, *Zimmerman* 1126!; Lindi District: Rondo Plateau, Mchinjiri, Jan. 1932, *Semsei* 622!

DISTR. **K** 7; **T** 3, 8; Mozambique (Vila Pery) and E. Zimbabwe

HAB. Evergreen forest; 350–850 m.

SYN. *A. zimmermannianus* Gilg in V.E. 3(2): 270 (1921), *nom. nud.*; T.T.C.L.: 555 (1949); Beentje, K.T.S.L.: 413 (1994), *nom. invalid.*

NOTE. Gilg's name has been used for this species but was never validated. I agree with pencilled comments by Vollesen on the covers suggesting the identity as *A. chirindensis.*

3. **A. dummeri** *Baker f.* in J.B. 57: 187 (1919); Radlk. in E.P. IV, 165: 551 (1932); I.T.U., ed. 2: 376 (1952); Hamilton, Field Guide Uganda For. Trees: 196 (1981). Type: Uganda, Mengo District, Kivuvu, *Dummer* 552 (BM!, holo., K!, iso.)

Shrub or spreading tree 2.5–9 m. tall; branches glabrous, lenticellate; slash red-brown exuding off-white sap. Leaves trifoliolate, drying bright green; petiole 4–14 cm. long; leaflets elliptic, elliptic-oblong or elliptic-obovate, 7.5–25 cm. long, 4–10.5 cm. wide, acuminate at the apex, cuneate at the base, entire or remotely coarsely serrate, thin to slightly thick, glabrous or with indistinct hairy domatia in nerve-axils beneath; lateral nerves in 5–12 pairs, venation fine and closely reticulate, raised on both sides. Inflorescence rather robust, 6–24.5 cm. long, including peduncles 5–8.5 cm. long, with 3–4 main branches; axes with short ferruginous pubescence, cymules 3–6-flowered, mostly densely placed. Flowers cream, white or greenish; calyx sparsely pubescent; outer sepals round, 1 mm. diameter; inner oblate, 2 mm. long and wide; petals spathulate, 2 mm. long, 0.8 mm. wide, ciliate. Fruits orange-yellow to red-brown, globose, ± 6 mm. diameter, with very sparse adpressed hairs or glabrous.

UGANDA. Bunyoro District: Budongo Forest, mahogany (Station III) sample plot, June 1940, *Eggeling* 3970! & Budongo Forest, 15 May 1941, *A.S. Thomas* 3893!; Mengo District: Mpanga Forest, 29 Jan. 1952, *Kaliebara* 43!

DISTR. **U** 2, 4; Zaire and Angola

HAB. Closed high forest, young forest, also in clearings and valley forest; 1050–1300 m.

SYN. *A. ussheri* Baker f. in J.B. 57: 186 (1919); Radlk. in E.P. IV, 165: 551 (1932). Type: Uganda, Mengo District, Kyagwe [Chagwe], Mabira Forest, *Ussher* 61 (K!, holo., BM!, iso.)
[*A. schweinfurthii* sensu Hauman in F.C.B. 9: 302 (1960) pro parte, *non* Gilg]

NOTE. The Uganda material is quite uniform and easily recognised. I am not convinced this is the same as *A. gossweileri* Baker f. described from Angola and after at first accepting this synonymy (suggested informally on the herbarium covers) have decided to revert to *A. dummeri* which, however, does also occur in Angola.

4. **A. africanus** *P. Beauv.*, Fl. Owar. 2: 75, t. 107 (1819); R.E. Fr., Wiss. Ergebn. Schwed. Rhod.-Kongo-Exped.: 131 (1914); Baker f. in J.B. 57: 188 (1919); V.E. 3(2): 270, fig. 131 (1921); Radlk. in E.P. IV, 165: 536, fig. 12 (1932); F.P.N.A. 1: 518 (1948); T.T.C.L.: 554 (1949); F.P.S. 2: 335 (1952); I.T.U., ed. 2: 375 (1952); C.F.A. 2: 77 (1954); F.W.T.A., ed. 2, 1: 713 (1958); K.T.S: 504 (1961); F.F.N.R.: 223 (1962); Haerdi in Acta Trop., suppl. 8: 122 (1964); Exell in F.Z. 2: 506 (1966); Exell & Sousa in Fl. Moçamb. 51: 13 (1973); Vollesen in Opera Bot. 59: 58 (1980); Troupin, Fl. Rwanda 2: 298, fig. 93/3 (1983); Vollesen in Fl. Ethiopia 3: 499, fig. 125.8/4 (1990); Beentje, K.T.S.L: 412, map (1994). Type: Nigeria, Oware, mouth of R. Formose, left bank, *Palisot de Beauvois* (G, holo., P-JU (no. 11367), iso.)

Shrub or tree (2–)4.5–10 m. tall, with heavy spreading crown; bark black or grey to reddish, smooth or rough; slash yellow with pink tinge; branchlets ± glabrous to puberulous or pubescent, lenticellate. Leaves trifoliolate, typically ± glabrous but hairy in variety, very discolorous, when glabrous dark and shiny above; leaflets narrowly elliptic to elliptic, 2–14.5 cm. long, 1–7 cm. wide, rounded to acuminate at the apex, attenuate to cuneate at the base, subentire to crenate or coarsely and irregularly dentate, subcoriaceous, sometimes with hairy domatia beneath; petiole (1–)2–7.5 cm. long. Inflorescences 4.5–15 cm. long, unbranched or usually with 1–6 branches, usually pale fulvous pubescent; cymules subsessile; pedicels (0.5–)1.5–2 mm. long, often glabrous. Flowers white, greenish white or yellowish green, scented; outer sepals broadly elliptic, ± 1 mm. long and wide, inner oblate, 1 mm. long, 1–1.5 mm. wide, ± glabrous; petal spathulate, 1 mm. long, 0.5 mm. wide. Filaments ± 2 mm. long, hairy at base. Style 1 mm. long, hairy; stigma 0.5 mm. long. Fruit ellipsoid, 5–7 mm. long, 4–4.5 mm. wide, very sparsely hairy. Seed ellipsoid, 5.5–6 mm. long, 4.5 mm. wide.

var. **africanus**; F.W.T.A., ed. 2, 1: 713 (1958), as forma

Leaves ± glabrous, dark and ± shiny above.

UGANDA. W. Nile District: Koboko, July 1938, *Hazel* 650!; Toro District: Katwe, Feb. 1933, *Hancock* 1304!; Mengo District: 50 km. on Kampala–Hoima road, Aug. 1932, *Eggeling* 482!

KENYA. Trans-Nzoia District: SE. Kitale, 24 Aug. 1986, *Beentje* 3007!; Kisumu-Londiani District: Muhoroni, 5 Apr. 1910, *Battiscombe* 289!; Masai District: Trans-Mara Plateau, Narok–Lolgorien road, 3.5 km. from Lolgorien, 15 Apr. 1961, *Glover, Gwynne & Samuel* 641!

TANZANIA. Bukoba District: Ruiga R. Forest Reserve, Apr. 1958, *Procter* 902!; Musoma District: Ndabaka Guard Post, 22 Feb. 1968, *Greenway, Kanuri & Myles Turner* 13329!; Mpanda District: Katavi [Katabi] Game Reserve, Sitalike [Starike], 5 km. on road to Katisunga, 20 Mar. 1973, *Ludanga* in *M.R.C.* 1545!

DISTR. **U** 1–4; **K** 3–6; **T** 1–8; widespread in Africa from Gambia to Cameroon, Zaire, Ethiopia and further south but taxonomy so confused that not possible to give distribution in detail

HAB. *Manilkara-Baikaea* forest, fringing forest (roots sometimes in water), forest margins, wooded and semi-open grassland, thicket and scrub, sometimes on termite mounds; (30–)650–1900(–2400) m.

SYN. *Schmidelia africana* (P. Beauv.) DC., Prodr. 1: 610 (1824) pro parte; Baker in F.T.A. 1: 421 (1868)

Allophylus schweinfurthii Gilg in E.J. 24: 286 (1897). Lectotype, chosen here: Sudan, *Schweinfurth* 3696 (K!, lecto.)*

A. goetzeanus Gilg in E.J. 28: 423 (1900); Radlk. in Sitz. Akad. Wiss. München 38: 220 (1909) & in E.P. IV, 165: 527 (1932); T.T.C.L.: 553 (1949). Type: Tanzania, Uhehe, Iringa, Kilima Plateau, *Goetze* 652 (B†, holo.)

A. subcoriaceus Baker f. in J.L.S. 37: 136 (1905); Radlk. in E.P. IV, 165: 547 (1932)*; I.T.U., 214 (1940); Hauman in F.C.B. 9: 293 (1960). Type: Uganda, Ankole, near Mulema, *Bagshawe* 264 (BM!, holo., K!, fragment)

A. africanus P. Beauv. forma *senegalensis* (A. Rich.) subforma *subcoriaceus* (Baker f.) Radlk. in E.P. IV, 165: 538 (1932)**

A. africanus P. Beauv. Group A sensu Exell in F.Z. 2: 507 (1966)

NOTE. The flowers are said to be attractive to flies; possibly many taxa in this genus are fly-pollinated.

Paulo 684 (Kenya, Masai District, Narok, Mosiro [Musilo], 19 Aug. 1960) a shrub to 5.5 m. in *Acacia* woodland and *Verdcourt* 1527 (Kenya, Masai District, 43 km. W. of Magadi, 'Sweet Waters', 29 July 1956) a liane to 7.5 m. with *Rhus natalensis, Cardiogyne, Ficus, Tamarindus*, etc., have thinner leaves much less shiny above giving a different facies but are referred here for the present.

* Gilg in E.J. 24: 286 (1897) cites *Schweinfurth* 3696 and 3668 from Sudan and *Preuss* 56 from Cameroon; I choose the K sheet bearing an original label of *Schweinfurth* 3696 as lectotype. This specimen is a variant of *A. africanus* sensu stricto.

** This dual naming of the same plant in the same work is doubtless the result that it was edited after Radlkofer's death.

The **T** 3 records are based on a few sheets from Pangani which are atypical and at low altitude.
Lynes Pr 58 (Dodoma District: 72–104 km. S. of Dodoma, plain of Little Ruaha R., 3 Mar. 1932) and *Lynes* Pr 40 (banks of Little Ruaha R., below Iringa, 22 Feb. 1932) agree well with Gilg's description of *A. goetzeanus* and cannot really be kept apart from typical *A. africanus*.
Much of the material from **T** 3, 6–8 has larger glabrescent leaflets much less or not shiny above. *A. elongatus* Radlk. in Sitz. Akad. Wiss. München 38: 221 (1909) & in E.P. IV, 165: 529 (1932); T.T.C.L.: 554 (1949), type: Tanzania, Lushoto District, W. Usambara Mts., Mlalo, *Holst* 288 (B†, holo.) may be this variant but the inflorescence is described as unbranched; it is referred to as *A. sp.* by Engler (Abh. Preuss. Akad. Wiss.: 60 (1894)).

var. **griseotomentosus** (*Gilg*) *Verdc.* comb. et stat. nov. Lectotype, chosen here: Tanzania, W. Usambara Mts., Lutindi, *Holst* 3426 (B†, syn.; K!, lecto.)*

Leaflets sparsely to densely grey pubescent beneath and often with hairy domatia in axils of main nerves and some veins as well.

UGANDA. Karamoja District: Kidepo National Park, 2 Sept. 1972, *Synnott* 1276!
TANZANIA. Ngara District: Bushubi, Nyakisasa, Kanyinya, 20 Dec. 1960, *Tanner* 5644!; Lushoto District: Korogwe, 16 Aug. 1969, *Archbold* 1083!; Morogoro District: Kimboza Forest Reserve, July 1952, *Semsei* 755!
DISTR. **U** 1; **T** 1–3, 5, 6, 8; Zaire, Zambia, Malawi, Mozambique and Zimbabwe (hairy variants also occur in West Africa and it is not possible to give detailed distribution)
HAB. Riverine forest, valley thicket; 200–1800 m.

SYN. *A. griseotomentosus* Gilg in E.J. 24: 290 (1897); Radlk. in Sitz. Akad. Wiss. München 38: 224 (1909); Baker f. in J.B. 57: 189 (1919); Radlk. in E.P. IV, 165: 540 (1932); T.T.C.L.: 554 (1949); Hauman in F.C.B. 9: 297 (1960); K.T.S.: 505 (1961)
A. griseotomentosus Gilg forma *genuinus* Radlk. in E.P. IV, 165: 540 (1932); T.T.C.L.: 554 (1949), *nom. illegit.* Type as for *A. griseotomentosus*
A. africanus P. Beauv. Group B sensu Exell in F.Z. 2: 507 (1966)

NOTE. Specimens with unbranched inflorescences merge with various forms of *A. rubifolius*.

5. **A. fulvotomentosus** *Gilg* in E.J. 24: 293 (1897); Z.A.E.: 476 (1912); Baker f. in J.B. 57: 189 (1919); Radlk. in E.P. IV, 165: 542 (1932); T.T.C.L.: 555 (1949); Hauman in F.C.B. 9: 296 (1960). Type: Tanzania, Mwanza District, Kagehi R., Utundua, *Stuhlmann* 3474 (B†, holo.)

Tree or shrub 1.8–7.5 m. tall, bark black or grey-brown, rough; stems usually densely ferruginous velvety, later ± glabrous. Leaves trifoliolate; petiole 1.5–6 cm. long, mostly ferruginous velvety; leaflets drying darker grey-brown above, pale brown beneath, elliptic to oblong or obovate, 2.5–12 cm. long, 1.5–8 cm. wide, subacute to rounded or ± truncate at the apex, cuneate to rounded at the base, mostly subcoriaceous, often densely ferruginous velvety at first, later hairy only on venation beneath or ± glabrous; venation reticulate and ± prominent above, the nerves ± impressed, very prominent beneath. Inflorescences 3.5–11(–15) cm. long, including 1–5 cm. long peduncle, branched, the axes ferruginous velvety; cymules subsessile, 3–10-flowered, extremely densely placed so that apices of inflorescence-branches are solid with flowers and spadix-like, the rhachis entirely obscured but often more open at base of branches and often altogether laxer in intermediates with *A. africanus*. Flowers yellow-green, subsessile or pedicels only 0.5 mm. long; inner sepals 1.5 mm. wide; petals spathulate, 1.5 mm. long, 0.8 mm. wide, the scale densely pilose. Filaments 2 mm. long, hairy at base. Fruits pale orange, subglobose-obovoid, 7 mm. long, 6 mm. wide, pubescent, subsessile.

* Other syntypes are *Stuhlmann* 8316 (**T** 6) and *Buchanan* 377, 472, 563, 573 & 800 (Malawi) all B†, but all Buchanan sheets save 563 are also at K.

TANZANIA. Ngara District: Bushubi, Muganza, Kumyangi, 3 Dec. 1959, *Tanner* 4597!; Mpwapwa District: Kiboriani Mt., 5 July 1938, *Mr. & Mrs. Hornby* 918!; Mbeya District: Mbosi, 29 Mar. 1932, *Clair-Thompson* 1075!

DISTR. **T** 1, 2, 4, 5, 7; Zaire and Burundi

HAB. Grassland with scattered trees, 'light forest', *Brachystegia* and *Combretum-Terminalia* woodland, thicket, cultivated steppe; 800–1800 m.

SYN. *A. kassneri* Baker f. in J.B. 57: 188 (1919); Radlk. in E.P. IV, 165: 552 (1932). Type: Zaire, Lufonzo, *Kassner* 2849 (BM!, holo., K!, iso.)

NOTE. The usually blunt leaflets, very dense inflorescences and often velvety indumentum give this a recognisable facies but there are intermediates with *A. africanus. Newbould & Jefford* 2230 (Mpanda District, Mahali Mts., Kalya) has ± glabrous leaflets truncate at the apex; together with several other similar sheets, I refer it here. Hauman gives the pedicel length as 1.5 mm. which is longer than found in the Flora area; he also records it from Uganda but I have seen no material. See also *A. rutete* on p. 97.

6. **A. pseudopaniculatus** *Baker f.* in J.L.S. 37: 137 (1905) & in J.B. 57: 188 (1919); Radlk. in E.P. IV, 165: 547 (1932). Type: Uganda, Ankole District, Rufua R., *Bagshawe* 544 (BM!, holo., K!, fragment)

Much-branched monoecious spreading tree or less often a shrub or climbing shrub, 3–15 m. tall; branches with brown lenticels, glabrous or glabrescent. Leaves drying very discolorous, dark purplish brown above, dark green beneath, trifoliolate; petiole 5–7 cm. long, glabrescent or ± tomentose; petiolule of terminal leaflet 1–3 mm. long, of laterals ± 1 mm. long; leaflets elliptic, oblong or obovate, 3–18 cm. long, 1.5–9 cm. wide, narrowly acutely acuminate at the apex, cuneate (or in laterals ± rounded and asymmetric) at the base, finely toothed, especially in the upper third; lateral nerves in ± 12 pairs. Inflorescences (5–)10–15 cm. long, simple or with 2 side branches up to 10 cm. long, axillary or appearing paniculate; peduncle 2–4 cm. long; axes with long hairs; cymules 4–7-flowered, the stalks ± 1 mm. long; pedicels 1–1.5 mm. long, glabrous. Flowers cream or yellow-green; outer sepals elliptic, inner oblate ± 1.8 mm. wide, glabrous or slightly pubescent; petals spathulate, 1 mm. long, the claw ciliate and scale covered with long hairs; disk lobed, glabrous. Stamens 2–2.5 mm. long, glabrous; staminodes 1.2 mm. long. Fruits green, of (1–)2–3 mericarps each subglobose, 5 mm. diameter, glabrous or with sparse hairs.

UGANDA. Ankole District: Rwabaranda Forest Reserve, 16 Feb. 1945, *Greenway & Eggeling* 7150!; Kigezi District: Rubuguli, Apr. 1948, *Purseglove* 2694!; Mengo District: Malubaga Forest, Aug. 1931, *Lab. Staff* S2182!

DISTR. **U** 1 (see note), 2, 4; **T** 1 (see note); E. Zaire and Rwanda

HAB. Swamp forest, valley bushland, hillside thicket and tangled termite mound vegetation; 1200–2250 m.

SYN. *A. kiwuensis* Gilg in Z.A.E.: 477 (1912); Baker f. in J.B. 57: 187 (1919); Radlk. in E.P. IV, 165: 548 (1933); Lebrun, Ess. For. Congo Orient.: 141, fig. 13B (1935); F.P.N.A. 1: 520, t. 50 (1948); Hauman in F.C.B. 9: 298, t. 31 (1960); Troupin, Fl. Rwanda 2: 298, fig. 93/1 (1983). Type: Zaire, Lake Kivu, Kwidjwi I., near Mgaturo, *Mildbraed* 1194 (B†, holo.)

A. crebriflorus Baker f. in J.B. 57: 187 (1919); Radlk. in E.P. IV, 165: 551 (1932); I.T.U., ed. 2: 376 (1952). Type: Uganda, Mengo, Kipayo, *Dummer* 680 (BM!, holo., K! (2 sheets), iso.)

NOTE. Gillett has annotated the cover of the Kew isotypes of *A. crebriflorus* suggesting they are *A. kiwuensis* and that the extremely dense inflorescence is pathological; although I can find no obvious proof of this no further material showing this character has turned up. *J. Wilson* 1584, Karamoja District, Nakiloro R.–Moroto Mt., may be a form of *A. kiwuensis* with leaflets densely almost velvety long-hairy beneath. *Tanner* 5290 (Tanzania, Ngara District, Bugufi, Mu Rgwanza) has similar leaflets and velvety ferruginous branchlets and may be another extreme variant.

7. **A. delicatulus** *Verdc.* sp. nov. a congeneribus inflorescentiis tenuibus, cymulis 1–2-floris, pedicellis tenuissimis 2–2.5 mm. longis distincta; *A. ferrugineo* Taub. similis inflorescentiis gracilioribus ramosis paucifloris differt. Typus: Tanzania, Ulanga District, Mahenge, Sali, *Schlieben* 1978 (BM!, holo.)

Shrub 2–4 m. tall, almost glabrous; bark pale grey or brownish grey. Leaves trifoliolate; petiole 3–6.5 cm. long, grooved, slightly pubescent at apex of junction of leaflets; petiolules ± absent, the lamina bases decurrent; leaflets drying bright green, elliptic or oblong-elliptic, 4–11 cm. long, 1.3–4 cm. wide, narrowly acutely acuminate at the apex, narrowly cuneate at the base, thin, serrate except at base, virtually glabrous; reticulation not raised, lateral nerves in 8–11 pairs. Inflorescences very slender, up to 16 cm. long, branched, the side branches few, ± 4 cm. long; peduncle up to 8 cm. long; cymule bases 0.5–3 mm. long, slender, 1–2-flowered; pedicels very slender, 2–2.5 mm. long; all axes virtually glabrous or micropuberulous. Outer sepals oblate, 1 mm. long, 1.8 mm. wide, glabrous or with few small marginal hairs; inner sepals round, convex, ± 1 mm. long and wide, petals obovate, 1 mm. long, the scale densely hairy. Filaments 1 mm. long, hairy at base; anthers 0.5 mm. long. Fruit not seen.

TANZANIA. Ulanga District: Mahenge, Sali, 24 Mar. 1932, *Schlieben* 1978!
DISTR. **T** 6; not known elsewhere
HAB. Mist forest; 1000–1100 m.

NOTE. On the same day in the same place Schlieben collected (1980) *A. ferrugineus* Taub.

8. **A. chaunostachys** *Gilg* in E.J. 30: 349 (1901); Radlk. in E.P. IV, 165: 524 (1932); T.T.C.L.: 553 (1949); Meikle in Mem. N.Y. Bot. Gard. 8: 240 (1953); Exell in F.Z. 2: 503, t. 100/6 (1966); Exell & Sousa in Fl. Moçamb. 51: 11 (1973); Brummitt in K.B. 31: 166 (1976); Troupin, Fl. Rwanda 2: 298, fig. 92/1 (1983). Type: Tanzania, Njombe District, Ukinga Mt., Yawiri Mt., *Goetze* 1196 (B†, holo., BM!, iso.)

Shrub, often straggling or scrambling or a distinct climber, less often a small tree (see note), (0.6–)1.5–4.5(–6) m. tall or long; stems very lenticellate, pubescent when young then glabrescent or ferruginous velvety. Leaves trifoliolate; petiole 2–7 cm. long, glabrous to densely pubescent; petiolules 0.2–1.5 cm. long or lateral leaflets often ± sessile; leaflets very discolorous, drying leaden grey above with the venation yellow and pale greenish or often bright brownish beneath, elliptic, oblong, rhombic or narrowly oblanceolate, 4–15.5 cm. long, 2–6 cm. wide, acute or acuminate at the apex, cuneate at the base, the laterals mostly very asymmetric, subentire to ± remotely serrate, crenate-serrate or coarsely serrate, glabrous to densely almost velvety ferruginous hairy and with tufts of hairs in axes of lateral nerves beneath; lateral nerves in 8–10 pairs, venation raised and reticulate above (impressed in life *fide* Milne-Redhead & Taylor) and strongly so beneath. Inflorescences unbranched or rarely branched, 5–17(–24) cm. long, including peduncle 1–8 cm. long, glabrous or spreading pubescent, the 1–few-flowered cymules usually well spaced; cymule stalk and rhachis 0.5–1 mm. long, pedicels 1–3 mm. long. Flowers usually ± large for the genus; outer sepals elliptic, 1.5 mm. long, 1 mm. wide, inner subcircular 1.5–1.8 mm. long, 1.5–3 mm. wide, glabrous or slightly pubescent; petals white or greenish yellow, spathulate, 1.5 mm. long, 1 mm. wide. Filaments 1.8 mm. long; staminodes 0.5 mm. long. Ovary 2-lobed, pubescent; style 2 mm. long. Fruits yellow or red, of 1–2 mericarps, each subglobose or obovoid, 6–6.5 mm. long, 5 mm. wide, minutely pubescent or glabrous.

TANZANIA. Ufipa District: above Sumbawanga, Mbizi Forest, 3 June 1980, *Hooper, Townsend & Mwasumbi* 1819!; Iringa District: Mufindi Tea estate, near Kibwele Forest, 17 Aug. 1984, *Bridson & Lovett* 545!; Njombe District: Nyumbanitu Forest, 21 July 1982, *Leliyo* 238!
DISTR. **T** 4, 7; E. Zaire, Rwanda, Burundi, Zambia, Malawi, Mozambique, Zimbabwe and South Africa (Gauteng)

HAB. Dense forest including relict patches, forest fringes, ravine forest, mist forest; (500–)1500–2250(–2400) m.

SYN. *A. yeru* Gilg in E.J. 30: 348 (1901); Radlk. in Sitz. Akad. Wiss. München 38: 221 (1909) & in E.P. IV, 165: 528 (1932); T.T.C.L.: 554 (1949). Type: Tanzania, Njombe District, Kondeland, Lumbila [Langenburg], Lumbila [Rumbira] valley, *Goetze* 832 (B†, holo., BM! (2 sheets), iso.)

A. buchananii Radlk. in Sitz. Akad. Wiss. München 38: 219 (1909); Baker f. in J.B. 57: 182 (1919); Radlk. in E.P. IV, 165: 524 (1932); Meikle in Mem. N.Y. Bot. Gard. 8: 240 (1953); Hauman in F.C.B. 9: 308 (1960); F.F.N.R.: 223 (1962). Type: Malawi, *Buchanan* 294 (B†, holo., BM, iso.)*

A. didymadenius Radlk. in Sitz. Akad. Wiss. München 38: 219 (1909) & in E.P. IV, 165: 525 (1932); Exell in F.Z. 2: 500 (1966). Type: Malawi, Masuku Plateau [Misuku Hills], *Whyte* (B†, holo., K!, iso.)

A. tenuifolius Radlk. in Sitz. Akad. Wiss. München 38: 221 (1909); Baker f. in J.B. 57: 184 (1919); Radlk. in E.P. IV, 165: 529 (1932). Type: Malawi, *Buchanan* 363 (B†, holo., BM!, K!, iso.)

A. gazensis Baker f. in J.B. 57: 182 (1919). Type: Zimbabwe, Chimanimani, *Swynnerton* 1321 (BM, holo., K!, SRGH!, iso.)

A. sp. near *buchananii* Radlk. sensu Meikle in Mem. N.Y. Bot. Gard. 8: 240 (1953)

A. sp. 1 sensu F.F.N.R.: 223 (1962)

A. uwembae Gilli in Ann. Naturhist. Mus. Wien 74: 441, t. 4/1 (1970). Type: Tanzania, Njombe District, Uwemba, *Gilli* 310 (W!, holo., K!, photo.)

NOTE. Exell (F.Z. 2: 504 (1966)) mentions that *Brass* 16802 from Mt. Mlanje was said to be a canopy tree to ± 30 m. and suspects the note was erroneous; Brass was, however, a very careful collector. Certainly none of the material now at Kew from Malawi exceeds 8 m. and mostly is much smaller and shrubby. In East Africa it merges with *A. ferrugineus* to some extent but has a distinct facies due to drying colours, strong venation and rather large flowers. *A. yeru* seems to be a small-flowered form of *A. chaunostachys.*

9. **A. ferrugineus** *Taub.* in P.O.A. C: 249 (1895); Z.A.E.: 475 (1912); Radlk. in E.P. IV, 165: 526 (1932); T.T.C.L.: 553 (1949); I.T.U., ed. 2: 376 (1952); Hauman in F.C.B. 9: 307 (1960); Beentje, K.T.S.L.: 412, map (1994). Type: Tanzania, Bukoba, *Stuhlmann* 1068 (B†, holo.)

Tree or shrub or sometimes a climber or creeper 0.9–9 m. long or tall, monoecious; bark grey, brown or reddish, smooth; slash white; young branchlets pubescent. Leaves trifoliolate; petiole (1–)2–10 cm. long; petiolule of terminal leaflet often well developed, 0.8–1.3(–2) cm. long; leaflets drying bright green, elliptic, ovate-oblong, rhomboid, obovate or oblanceolate, (2.5–) 6–22 cm. long, (0.9–)2.5–11 cm. wide, acute to acuminate at the apex, attenuate to cuneate at the base, the laterals very asymmetric, regularly toothed or serrate, glabrous, pubescent, ferruginous hairy or velvety, with or without hairy domatia in the axils of nerves beneath; lateral nerves in 5–10 pairs. Inflorescences (3.5–)7.5–34(–47) cm. long (see note), usually unbranched, often slender and drooping; peduncles 1.5–5 cm. long; cymules 1–6-flowered, the stalks 1.5–5 mm. long, usually well-spaced; pedicels 1.5–3(–5) mm. long, minutely puberulous. Flowers green, white, greenish white or yellow; outer sepals elliptic, 1–1.5 mm. long, 1 mm. wide, sparsely puberulous, inner ± round, 1.5–2 mm. diameter; petals spathulate, 2 mm. long, 1 mm. wide. Filaments ± 2 mm. long, hairy in basal half. Style ± 1 mm. long, with few hairs; stigma 0.5–1 mm. long. Fruit orange to scarlet, often of 2–3 mericarps, each ellipsoid, obovoid or globose, 5–8 mm. long., 4–6.5 mm. wide, or 7.5 mm. diameter, at first sparsely hairy, glabrescent.

* One of the many specimens sent as 'duplicates' from Kew at that time which were assumed to be common and not worth retaining.

var. **ferrugineus**

Cymules not noticeably elongate, the rhachis 1.5–2 mm. long.

UGANDA. Kigezi District: Maramagambo [Malamagambo] Forest, Feb. 1950, *Purseglove* 3290!; Mbale District: Bugisu [Bugishu], Bufumbo, Nov. 1932, *Chandler* 1011!; Mengo District: Mityana, Kasa Forest, 17 Jan. 1950, *Dawkins* 489!

KENYA. Elgon, 20 July 1931, *Mrs. C. Lugard* 671!; Meru District: Nyambeni Tea Estate, 8 Oct. 1960, *Verdcourt & Polhill* 2939!; N. Kavirondo District: Kakamega Forest, June 1961, *Lucas* 139!

TANZANIA. Moshi District: Kilimanjaro, N. of Mweka, 11 Apr. 1968, *Renvoize & Abdallah* 1413!; Morogoro District: Bunduki, Kimbinyuko, Mar. 1953, *Semsei* 1115!; Iringa District: Mwanihana Forest Reserve, above Sanje, 10 Oct. 1984, *D.W. Thomas* 3292!

DISTR. **U** 2–4; **K** 3–6; **T** 1, 2, 4, 6, 7; Cameroon, Central African Republic, Zaire, Rwanda, Sudan, Ethiopia and Angola

HAB. Forest, often by streamsides, hillside thicket, sometimes on termite mounds, plantations; (650–)1050–2400(–?2700) m.

SYN. *A. macrobotrys* Gilg in E.J. 24: 288 (1897); Radlk. in E.P. IV, 165: 524 (1932); F.P.N.A. 1: 517 (1948); T.T.C.L.: 553 (1949); I.T.U., ed. 2: 376 (1952); Hauman in F.C.B. 9: 305 (1960); K.T.S.: 505 (1961); Hamilton, Field Guide Uganda For. Trees: 196, fig. 352 (1981); Troupin, Fl. Rwanda 2: 300, fig. 92/2 (1983); Vollesen in Fl. Ethiopia 3: 499, fig. 125.8/8–9 (1990). Type: Tanzania, Bukoba, *Stuhlmann* 3315, 3598, 3759 (B†, syn.)*

A. macrurus Gilg in E.J. 24: 287 (1897); Radlk. in Sitz. Akad. Wiss. München 38: 219 (1909) & in E.P. IV, 165: 525 (1932). Type: Zaire, between Balaibo and Daki, *Stuhlmann* 2783 (B†, holo., BM!, fragment)

A. oreophilus Gilg in E.J. 24: 289 (1897); Radlk. in Sitz. Akad. Wiss. München 38: 219 (1909); Z.A.E.: 474 (1912); R.E. Fr., Wiss. Ergebn. Schwed. Rhod.-Kongo-Exped.: 131 (1914); Baker f. in J.B. 57: 182 (1919); Radlk. in E.P. IV, 165: 525 (1932); F.P.N.A. 1: 516 (1948); Hauman in F.C.B. 9: 307 (1960); K.T.S.: 505 (1961). Type: Uganda, Toro District, Ruwenzori [Runssoro], Ruimi [Wimi], *Scott Elliot* 7938 (B†, holo., K!, iso.)

A. schirensis Gilg in E.J. 24: 289 (1897); Radlk. in Sitz. Akad. Wiss. München 38: 223 (1909), as '*shirensis*', & in E.P. IV, 165: 533 (1932), as '*shirensis*'; T.T.C.L.: 554 (1949). Type: Tanzania, Kilimanjaro, Schira, path from Kibo, *Volkens* 1937 (B†, holo.)

A. volkensii Gilg in E.J. 24: 290 (1897); Radlk. in Sitz. Akad. Wiss. München 38: 219 (1909) & in E.P. IV, 165: 526 (1932); T.T.C.L.: 553 (1949); K.T.S.: 506 (1961). Type: Tanzania, Kilimanjaro, Marangu, *Volkens* 2077 (B†, holo.)

A. welwitschii Gilg in E.J. 24: 287 (1897); Radlk. in Sitz. Akad. Wiss. München 38: 222 (1909); Baker f. in J.B. 57: 184 (1919); Radlk. in E.P. IV, 165: 530 (1932); F.P.S. 2: 335 (1952); C.F.A. 2: 76 (1954); F.W.T.A., ed. 2, 1: 714 (1958); Hauman in F.C.B. 9: 310 (1960); Fouilloy & Hallé in Fl. Cameroun, 16: 34, t. 5/1–4 (1973) & in Fl. Gabon 23: 34, t. 5/1–4 (1973). Type: Angola, Golungo Alto, Sobado, Quilombo-Quiacatubia, *Welwitsch* 4510 (B†, holo., BM!, COI, K!, LISU, iso.)

A. latefoliolatus Baker f. in J.L.S. 37: 137 (1905); Radlk. in Sitz. Akad. Wiss. München 38: 226 (1909); Baker f. in J.B. 57: 182 (1919); Radlk. in E.P. IV, 165: 546 (1932). Type: Uganda, Mengo District, Lake Victoria, Musozi, *Bagshawe* 153 (BM!, holo.)

?*A. brachycalyx* Baker f. in J.B. 57: 188 (1919); Radlk. in E.P. IV, 165: 552 (1932). Type: Uganda, Toro/Mubende District, Lake Albert, near Muzizi [Mizizi], *Bagshawe* 1325 (BM!, holo.)

A. cuneatus Baker f. in J.B. 57: 183 (1919); Radlk. in E.P. IV, 165: 550 (1932); K.T.S.: 505 (1961). Type: Kenya, Kiambu District, Limuru, *Dummer* 1566 (BM!, holo., K!, iso.)

A. toroensis Baker f. in J.B. 57: 181 (1919); Radlk. in E.P. IV, 165: 549 (1932). Types: Uganda, Toro District, near Mpanga R., *Bagshawe* 1128 & near mouth of Mpanga R., *Bagshawe* 1150 (BM!, syn.)

A. sp. sensu K.T.S.: 506 (1961) quoad *Napier* 2930 in *C.M.* 5340

NOTE. *A.S. Thomas* 54 (Uganda, Masaka District, Sese Is., Bugala, Mugoye, 4 June 1932) is clearly this species but has long inflorescences with 1–3 basal branches. *Maitland* 515 (Uganda, Mengo District, Mabira Forest, Dec. 1922) has abnormal inflorescences with pedicels ± 5–6 mm. long. *E.J. & C. Lugard* 508 (Kenya, Elgon, 12 Jan. 1931) has an extraordinary multibranched abnormal (galled?) inflorescence. A sterile specimen *S.A.*

* Radlkofer has labelled *Dawe* 136 (Uganda, Mubende District, Bugangazi [Bugangadzi]) as 'vix dubie *Allophylus macrobotrys* Gilg'.

Robertson & Luke 5515 (Kenya, Lamu District, Witu Forest, 11 Nov. 1988 at 18 m.) appears close – nothing else has been seen from **K** 7 nor at low altitude. Mt. Hanang (Mbulu District) populations have been variously referred to *A. stachyanthus* or *A. rubifolius*, e.g. *Greenway* 7732 (Hemit, 2225 m.) and *Polhill & Paulo* 2298 (above Katesh, 1980 m.). I have referred them to *A. ferrugineus* but am far from certain about the correctness of this treatment.

var. **stipitatus** *Verdc.*, var. nov. a var. *ferrugineo* rhachidi cymularum elongata, 4–5 mm. longa differt. Typus: Tanzania, Mpanda District, Kungwe Mt., Ntale R., *Newbould & Harley* 4720 (K!, holo.)

Shrub 1.8–3 m. or small tree 6 m. tall, sometimes branched only at apex. Leaflets up to 22 cm. long, 11 cm. wide, ± glabrous. Inflorescences 9–47 cm. long; cymules with rhachis elongate, 4–5 mm. long. Fruits 8–9 mm. in diameter.

TANZANIA. Mpanda District: Mahali Mts., Kasangazi–Ujamba footpath, 18 July 1958, *Mgaza* 166!; ?Kigoma District: without locality, July 1958, *Mahinda* 100B! (sterile); Morogoro District: Bunduki, Kimbinyuko, Mar. 1953, *Semsei* 1113!
DISTR. **T** 2 (Rau), 4, 6, 7; not known elsewhere
HAB. Steep forested ravines, forest edges; 1000–2100 m.

NOTE. A few sheets from elsewhere come close, e.g. *Polhill* 361 (Kiambu District, Limuru, 26 Mar. 1961).

10. **A. sp.**

Shrub to 1 m. tall; branchlets sparsely adpressed-pubescent but soon glabrous. Leaves drying pale green; petioles 2.5–6.5 mm. long, shortly pubescent; leaflets ovate, 5–11 cm. long, 3–6 cm. wide, acuminate at the apex, cuneate at the base, the laterals distinctly asymmetric, rather regularly shortly toothed, sparsely adpressed-pubescent becoming ± glabrous. Inflorescences 11–17 cm. long, with some branches held at ± right angles 2.5–4.5 cm. long; cymules few-flowered with stalks 2 mm. long; pedicels 3–5 mm. long.

UGANDA. Toro District: Ruwenzori, Nyamwamba [Namwamba] valley, Kilembe, 21 Dec. 1934, *G. Taylor* 2584!
DISTR. **U** 2; not known elsewhere
HAB. Probably forest, near bamboo boundary; 2700 m.

NOTE. The inflorescences are to some extent pathological with a few flowers developing a pair of small simple leaves, instead of style-arms and stigmas; nevertheless it seems unlikely that the distinctively regularly branched inflorescences are abnormal but if so then it is probably only a curious form of *A. ferrugineus*. As mentioned under that species obviously abnormal inflorescences with extensive branching are known. No other material has been seen. No habitat data are given other than altitude; Kilembe itself is 1200 m. but the path up the valley quickly reaches over 2700 m.

11. **A. tanzaniensis** *F.G. Davies* sp. nov., affinis *A. rubifolii* (Hochst.) Engl. var. *alnifolii* (Baker) Radlk., foliolis oblongis ellipticis vel oblanceolatis haud rotundo-obovatis, fere glabris, inferne subintegris, ± acuminatis, petiolis (2–)5–9 cm. longis, inflorescentibus glabris vel minute puberulis, etiam *A. africano* P. Beauv. similis, inflorescentiis simplicibus vel parce ramosis glabris vel puberulis differt. Typus: Tanzania, Lindi District, Lake Lutamba, *Schlieben* 5895 (K!, holo., BM!, K!, iso.)

Much-branched shrub or small tree, 1–5 m. tall; branches pale to medium brown with brown lenticels; youngest shoots pubescent, soon glabrous. Leaves drying pale green or ± darker above, trifoliolate; petiole (2–)5–9 cm. long, slender, puberulous or glabrous; petiolule of terminal leaflet 2–7 mm. long, of lateral 0–3 mm. long; leaflets discolorous, drying brown or brownish green above, subequal or laterals $^1/_2$ to $^4/_5$ length of the terminal one, elliptic or elliptic-oblong, (1.5–)5–9 cm. long,

(0.9–)2.3–4.2 cm. wide, ± rounded to shortly obtusely acuminate at apex, terminal one cuneate at the base, laterals asymmetrically rounded, entire or crenate or toothed for upper $^1/_2$ or $^2/_3$, ± thin, glabrous or with few hairs on nerves beneath; hairy domatia sometimes present; lateral nerves in ± 7 pairs, rather inconspicuous. Inflorescences slender, (3–)8–12(–17) cm. long inclusive, unbranched or with 1–2 short branches, the axes quite glabrous, puberulous or in some specimens pubescent; cymules 3–9-flowered, somewhat spaced, subsessile; peduncles (1–)2.5–5.5 cm. long, pedicels 1–1.8 mm. long. Flowers white or yellow; sepals ± round, outer 1 mm. diameter, inner 1.2–1.5 mm. diameter; petals spathulate, 1 mm. long, 0.5–0.8 mm. wide, long white-ciliate at base. Filaments 1.5 mm. long, glabrous. Fruit red, globose, 5.5 mm. diameter.

TANZANIA. Dodoma District: Kigwe, 20 May 1978, *Sigara* in *Ruffo* 1165!; Kilosa District: 3–10 km. on Malolo–Kisanga track, 3 Apr. 1988, *Bidgood, Mwasumbi & Vollesen* 901!; Mbeya District: Songwe R., 27 Jan. 1963, *Richards* 17557!

DISTR. **T** 5–8; not known elsewhere but probably in Mozambique

HAB. Deciduous bushland, *Acacia, Premna, Commiphora, Euphorbia* 'light woodland', 'Makonde thicket' on orange sand, sometimes on limestone, rocky hillsides etc.; 350–1400(–?1560) m.

SYN. *A. sp. nov.* (*sp. nov.* 2 of Kew) sensu Vollesen in Opera Bot. 59: 58 (1980)

NOTE. Although the material 'hangs together' and according to workers who have seen it in the field it has a distinctive facies, it is very difficult to separate in words from other species; it is altogether more glabrescent than most material of *A. ferrugineus, A. rubifolius* and *A. africanus,* the latter to which some specimens with slightly branched inflorescences might be referred; *Richards* 17557 cited above has 1–2 short branches in the lower part of the inflorescence.

12. **A. rubifolius** (*A. Rich.*) *Engl.* in Hochgebirgsfl. Trop. Afr.: 292 (1892); Baker f. in J.B. 57: 159 (1919); Radlk. in E.P. IV, 165: 520 (1932); T.T.C.L.: 553 (1949); F.P.S. 2: 335 (1952); Hauman in F.C.B. 9: 305 (1960); K.T.S.: 506 (1961); Exell in F.Z. 2: 502, t. 100/5 (1966); Exell & Sousa in Fl. Moçamb. 51: 8 (1973); Vollesen in Opera. Bot. 59: 58 (1980); Troupin, Fl. Rwanda 2: 300, fig. 93/2 (1983); Vollesen in Fl. Ethiopia 3: 502, fig. 125.8/1–3 (1990); Beentje, K.T.S.L.: 413, map (1994). Type: Ethiopia, Tigre, Djeladjeranne, *Schimper* III. 1613 (P, syn., BM!, FI (Webb), K!, isosyn.) & Tacazze valley, *Schimper* III. 1669 (P, syn., FI (Webb), BM!, K!, UPS, isosyn.)

Shrub or small to medium-sized tree 3–7(–12) m. tall, also occasionally described as a creeper; bark grey, smooth or rough; branchlets glabrous to sparsely pubescent or subtomentose. Leaves trifoliolate, mostly discolorous and drying deep or purplish brown above, grey-green beneath; petiole 0.8–5(–6.5) cm. long; leaflets elliptic, distinctly obovate or ± round, 1–10(–14) cm. long, 0.8–5.5(–8) cm. wide but predominantly small, emarginate to acute (mostly rather obtuse) at the apex, attenuate to cuneate at the base, shallowly crenate to irregularly dentate or serrate, glabrous to densely pubescent or velvety and often with conspicuous hair-tufts in nerve-axils beneath; indumentum usually greyish. Inflorescences usually unbranched, 2–18(–22) cm. long, but occasionally with a few branches 1–3(–12) cm. long; peduncles 1.2–5.5 cm. long; cymules 1–5-flowered, subsessile, usually closely placed, sometimes very dense; pedicels 1–2(–3) mm. long, pubescent or glabrous. Flowers green or yellowish white; outer sepals broadly elliptic, ± 1 mm. long and wide, inner ± round, 1 mm. long, 1–1.5 mm. wide; petals spathulate, 1 mm. long, 0.5 mm. wide. Filaments 1–2 mm. long, hairy at base. Style 1 mm. long, hairy; stigma 0.5 mm. long. Fruits orange to bright red, of 1–2 mericarps, each ellipsoid, obovoid or subglobose, 6–8 mm. long, 5–6 mm. wide, finely pilose.

The key beneath is only the roughest of guides.

KEY TO INFRASPECIFIC VARIANTS

1. Leaves essentially glabrous 2
   Leaves pubescent to velvety 3
2. Leaflets rounded obovate, usually small, up to 3.5 × 2.8 cm., mostly rounded at apex . . . . c. var. **alnifolius** (the typical coastal form)
   Leaflets mostly elliptic, often quite large up to 14 × 8 cm., often acute or acuminate . . . . forms of var. **alnifolius** and other varieties of *A. rubifolius* and intermediates with *A. ferrugineus*
3. Leaflets elliptic, mostly 7–10 × 2–5.5 cm., thickly velvety beneath; inflorescences up to 17 cm. long, often with cymules very dense, sometimes branched . . . b. var. **dasystachys**
   Leaflets obovate and smaller, pubescent to ± velvety beneath; inflorescences shorter and with cymules laxer . . . . a. var. **rubifolius**

a. var. **rubifolius**

Leaflets sparsely to densely pubescent, typically obovate. Inflorescence (2.5–)5–10(–12) cm. long.

UGANDA. Karamoja District: E. of Loyoro, towards the Turkana Escarpment, July 1954, *M.S. Philip* 640!; Bunyoro District: Butiaba Flats, Oct. 1933, *Eggeling* 1437!; Mbale District: Bukedi, 3 km. N. of Malaba R., near Apoli, 5 July 1971, *Katende* 1144!

KENYA. Northern Frontier Province: S. end Mt. Kulal, Mar. 1972, *Tweedie* 4291!; E. Elgon, June 1960, *Tweedie* 2016B!; Kitui, Kyangwithie, 20 Jan. 1962, *Mbonge* 11!

TANZANIA. Musoma District: Serengeti, Seronera, 9 Jan. 1963, *Greenway* 10932!; Kondoa District: Farkwa, 27 Jan. 1966, *Newman* 123!; Uzaramo District: Dar es Salaam, University College, 22 Feb. 1969, *B.J. Harris & Gardiner* 2806!

DISTR. **U** 1–4; **K** 1–7; **T** 1–8; E. Zaire, Sudan, Ethiopia and N. Somalia, south to South Africa (Gauteng and Kwazulu-Natal)

HAB. Grassland with scattered trees, rough grassland, thicket, edges of cultivations, woodland, sometimes riverine; ?0–2250 m.

SYN. *Schmidelia rubifolia* A. Rich., Tent. Fl. Abyss. 1: 103 (1847); Baker in F.T.A. 1: 423 (1868)
A. fischeri Gilg in E.J. 24: 292 (1897); Type: Kenya/Tanzania, probably 'Massai Steppe', *Fischer* 1: 308 (B†, holo.)
A. tristis Radlk. in Sitz. Akad. Wiss. München 38: 225, 237 (1909); Baker f. in J.B. 57: 188 (1919); Radlk. in E.P. IV, 165: 542 (1932). Type: Mozambique, Sena, Chirumbe and Chupanga, *Kirk* (K!, syn.) & Quelimane, *Stuhlmann* 670 (B†, syn.) & Puguruni, *Stuhlmann* 670 (B†, syn.)
A. erlangeri Chiov. (1932), publication not traced; E.P.A.: 491 (1958). Types: Ethiopia, S. Harar, between Gallaborda and Sheik Hussein, *Ellenbeck** (B†, syn.) & Laga Hida, *Basile* 330 (TO, holo.)
[*A. alnifolius* sensu Radlk. in E.P. IV, 165: 464, 521 (1932) pro parte; E.P.A.: 491 (1958) pro parte, *non* (Baker) Radlk. sensu stricto – Radlkofer restricted *A. rubifolius* to Ethiopia]
?*A. africanus* Groups B & C sensu Exell in F.Z. 2: 507 (1966) pro parte – with simple inflorescences
[*A. griseotomentosus* sensu adnot. mult., *non* Gilg sensu stricto]

NOTE. Radlkofer (E.P. IV, 165: 522 (1932)) synonymises *A. fischeri* with *A. alnifolius* but Gilg's original description mentions that the leaflets are densely grey tomentose beneath and small; it belongs here I think.

* Index Kewensis gives "Kenya Colony" for this specimen (under *A. erlangeri* Gilg ex Engl., *nom. nud.*).

b. var. **dasystachys** (*Gilg*) *Verdc.* comb. et stat. nov. Type: Kenya, Nairobi/Masai Districts, Athi Plains, Feb. 1896, *Pospischil* (W, holo., K!, photo.)

Leaflets very discolorous, drying dark brown above, pale grey brown thickly velvety beneath, usually more elliptic and 7–10(–13) cm. long, 2–5.5(–7.5) cm. wide. Inflorescences up to 17 cm. long, often with cymules very dense, sometimes branched (see note).

KENYA. Nairobi District: Bahati, Mbagathi stream, 27 Mar. 1933, *Gilbert Rogers* 545! & Karura Forest, 26 Mar. 1958, *Verdcourt* 2143!; Machakos District: Ulu, 6 Mar. 1930, *Napier* 22!

TANZANIA. Masai District: Namanga R., 28 June 1938, *Evans & Erens* 1054!; Arusha District: Duluti, Tengeru, 5 July 1954, *Matalu* 3154!; Pare District: SW. Pare Mts., near Vudee, 30 Jan. 1930, *Greenway* 2081!

DISTR. **K** 1, 4, 6, 7; **T** 1–3; not known elsewhere

HAB. Dry evergreen forest, open woodland, bushland, cultivations, etc.; (50–150 (coast)) 900–2100 m.

SYN. [*Schmidelia rubifolia* sensu Oliv. in Trans. Linn. Soc., ser. 2, 2: 331 (1887), *non* A. Rich.]

[*Allophylus rubifolius* sensu Engl., Hochgebirgsfl. Trop. Afr.: 292 (1892) pro parte, *non* (A. Rich.) Engl. sensu stricto]

A. dasystachys Gilg in E.J. 24: 293 (1897); Radlk. in Sitz. Akad. Wiss. München 38: 220 (1909) & in E.P. IV, 165: 526 (1932)

A. stachyanthus Gilg in E.J. 24: 292 (1897); Radlk. in Sitz. Akad. Wiss. München 38: 225 (1909) & in E.P. IV, 165: 541 (1932); T.T.C.L.: 555 (1949); K.T.S.: 506 (1961). Types: Kenya, Teita District, Mwatate, *Johnston* (B†, syn., BM!, K!, isosyn.) & Tanzania, Kilimanjaro, Marangu, *Volkens* 618 & 1703 (B†, syn.) & Pare District, Pare Mts., Ugweno [Ugueno] Mt., *Volkens* 495 (B†, syn.) & Biharamulo District, Bukome, *Stuhlmann* 3460 (B†, syn.)

A. calophyllus Gilg in E.J. 24: 291 (1897); Radlk. in Sitz. Akad. Wiss. München 38: 224 (1909) & in E.P. IV, 165: 541 (1932); T.T.C.L.: 555 (1949). Type: Tanzania, Kilimanjaro, Useri, *Volkens* 1973 (B†, holo.)

[*A. spragueanus* sensu T.T.C.L.: 553 (1949) quoad *Greenway* 2081, *non* Burtt Davy* sensu stricto]

NOTE. Around Nairobi this is a stable entity with exactly similar specimens having been collected in the same place over a period of 25 years but forms occur with leaves with only a few scattered hairs, e.g. *Mettam* 164 (Kiambu District, Kabete, 1860 m.) - it is by inspection clearly the same taxon. The dense-flowered forms are distinctive, e.g. *Battiscombe* 934 which has many more male than female flowers (Nairobi Arboretum - wild part, c. 1919). Coastal forms usually have much shorter inflorescences. Forms with branched inflorescences are rather frequent and correspond to *A. stachyanthus*. It is frequently impossible to name them and many have been determined as hybrids with *A. africanus*, but on no evidence.

var. **alnifolius** (*Baker*) *Friis & Vollesen* in K.B. 51: 802 (1996). Type: Mozambique, locality uncertain, *Forbes* (K!, holo., BM!, K!, iso.)

Mature leaves almost entirely glabrous save for scattered hairs on venation; leaflets in common coastal form small, 0.9–3.5 cm. long, 0.7–2.8 cm. wide, rounded-obovate, mostly emarginate to rounded at the apex, but longer-leaved forms have also been referred here with leaflets up to 5.5–7.5 × 3–4 cm. and acute at the apex. Inflorescences typically short, 4–5 cm. long, but much longer in some forms.

UGANDA. Karamoja District: Nabilatuk to Lolacat (?Loliakat), 15 June 1957, *Dyson-Hudson* 205! (intermediate)

KENYA. Kitui, May 1877, *Hildebrandt* 2872!; Teita District: 80 km. on Mombasa–Voi road, Nov. 1937, *Dale* 3760!; Tana River District: 50 km. S. of Garsen, Kurawa, 25 Sept. 1961, *Polhill & Paulo* 576!

TANZANIA. Pangani District: Mkwaja, Mkaramo, 8 Jan. 1957, *Tanner* 3373!; Bagamoyo District: Kikoka Forest Reserve, 6 Mar. 1966, *Mgaza* 848!; Uzaramo District: Kisarawe, Banda Forest Reserve, Dec. 1964, *Procter* 2809!; Zanzibar, *Toms* 26!

DISTR. **U** 1; **K** 1–5, 7; **T** 2, 3, 6; **Z**; Ethiopia, Somalia and Mozambique

* *A. spragueanus* Burtt Davy in K.B. 1921: 280 (1921); Radlk. in E.P. IV, 165: 1485 (1934). Type: Zimbabwe, Victoria Falls, *Allen* 143 (K, holo.!)

HAB. Evergreen thicket often on coral rag, dry forest edges, *Commiphora* and *Diospyros-Acacia* etc. bushland, grassland with scattered trees; 0–700 m., but intermediates in *Juniperus* forest etc. at 1000–1400(–1900) m.

SYN. *Schmidelia alnifolia* Baker in F.T.A. 1: 422 (1868)
S. repanda Baker in F.T.A. 1: 422 (1868). Types: Mozambique, R. Rovuma, *Kirk* & Lower Chire [Shire] valley, *Kirk* & Chire [Shire], Elephant Marsh, *Meller* (all K!, syn.)
Allophylus repandus (Baker) Engl. in E.J. 17: 160 (1893); Baker f. in J.B. 57: 185 (1919)
A. alnifolius (Baker) Radlk. in E. & P. Pf. III, 5: 313 (1895); Baker f. in J.B. 57: 159 (1919); Radlk. in E.P. IV, 165: 521 (1932); Chiov., Fl. Somala 2: 145 (1932); T.T.C.L.: 553 (1949); Hauman in F.C.B. 9: 304 (1960); K.T.S.: 504 (1961); Exell in F.Z. 2: 501, t. 100/7 (1966); Exell & Sousa in Fl. Moçamb. 51: 7 (1973); Vollesen in Fl. Ethiopia 3: 502, adnot. (1990)
A. kilimandscharicus Taub. in P.O.A. C: 249 (1895); Radlk. in Sitz. Akad. Wiss. München 38: 218 (1909) & in E.P. IV, 165: 519 (1932); T.T.C.L.: 553 (1949); Beentje, K.T.S.L.: 413, map (1994). Type: Tanzania, Kilimanjaro, Useri, in the Tarakia [Karrakia] gorge, *Volkens* 2003 (B†, holo.) – Radlkofer also added *Endlich* 148 from Ol Molog in his later reference (see notes)
Schmidelia minutiflora Mattei in Boll. Ort. Bot. Palermo 7: 179 (1908) & in F.R. 9: 346 (1911); Baker f. in J.B. 57: 159 adnot. (1919). Type: Ethiopia, Giumbo, Goscia, *Macaluso* (PAL, holo.) (e descr.)
Allophylus tenuis Radlk. in Sitz. Akad. Wiss. München 38: 218 (1909) & in E.P. IV, 165: 520 (1932); K.T.S.: 506 (1961). Type: Kenya, Teita District, Ndi, *Hildebrandt* 2562 (B†, holo., BM!, K!, iso.)

GENERAL NOTES ON *A. rubifolius*. There remain a large number of forms of which there may be large stable populations in the areas concerned. Glabrous- or almost glabrous-leaved forms have mostly been labelled as var. *alnifolius* but are very different from the small-leaved coastal form traditionally looked on as typical *alnifolius* (but not by any means identical with the type), e.g. *Newbould* 7398 (**K** 1, Northern Frontier Province, upper Turkwell R., Kaputir, 900 m.) has obovate ± glabrous leaflets, up to 9 × 5.5 cm. These are perhaps glabrous forms of var. *dasystachys* but also approach *A. ferrugineus*. A similar form is from Zanzibar, *Faulkner* 2388 (Mkokotoni, 26 Oct. 1959) with glabrescent leaflets 10 × 5 cm. and very densely flowered inflorescences 1–23 cm. long, and there is actually a name based on this form - *A. stachyanthus* Gilg forma *calvus* Radlk. in E.P. IV, 165: 542 (1932); type: Zanzibar, [Mkokotoni] Kokotoni, *Stuhlmann* 703 & 704 (B†, syn.). Other similar forms but with branched inflorescences have been referred to *A. africanus*, e.g. *Paulo* 684 (Masai District, Mosiro [Musilo], 19 Aug. 1960) and *Verdcourt* 1527, a liane to 7.5 m. (Masai District, 43 km. W. of Magadi, 'Sweet Waters', 29 July 1956).

A form from **K** 2, 3 (Baringo) with glabrescent leaflets, coarsely crenate in upper half and cuneate and entire beneath has somewhat the facies of *A. tanzaniensis* F.G. Davies, e.g. *Hemming* 361 (Turkana District, 40 km. W. of Kangetet, 7 July 1954, in sandy riverbed); it has been considered a form of var. *alnifolius*.

Around Nairobi (1700–1950 m.) there is a well known form which I have referred to var. *alnifolius* which has usually been called *A. aff. repandus* (Baker) Radlk. or *A. kilimandscharicus* Taub.; it has glabrous (save for domatia) coarsely toothed leaves and somewhat larger flowers. Populations have certainly been constant for over a century as evidenced by the numerous collections, e.g. *Bally* 10504, *Dale* 2884, *Dummer* 2007, *Kibue* 157, *Linton* 135, *Mearns* 179, *Napier* 294, *Polhill* 357, *Rayner* 18, *Rogers* 469, 547 & 548, and *Whyte*. There are, however, too many intermediates when a much larger area is considered but local botanists could call it var. *alnifolius* 'Nairobi form'.

I am not entirely certain of the identity of the fragment of the type of *A. kilimandscharicus*. It could be an abnormally small leaved form of *A. ferrugineus*.

Something must also be said about forms occurring on Zanzibar. The Mkokotoni form has been mentioned above and var. *alnifolius* certainly occurs but other material is more difficult. *Vaughan* 1065 (Mbweni, 9 Jan. 1930) has been referred to *A. rubifolius* var. *rubifolius* but is only sparsely hairy and intermediate with var. *alnifolius*; the inflorescences are sparsely branched which has led to it being named *A. africanus*. *Faulkner* 3359 (Chukwani, 14 Mar. 1964) and *Vaughan* 1421 (Chukwani, 28 July 1930) with simple inflorescences are similar and the same kind of intermediates. A really detailed study of the populations on the island might help one to understand the reproductive biology of the genus.

Gereau et al. 4997 (Ufipa District, 8–9 km. from Sumbawanga on Mbala road, Safu, in *Brachystegia* woodland at 1600 m.) with stiff branched inflorescences from leafless nodes, obovate leaflets 6.5 × 4 cm., with rounded apices and yellow-grey indumentum may be better considered a form of *A. africanus*.

13. **A. richardsiae** *Exell* in Bol. Soc. Brot., sér 2, 38: 109 (1964) & in F.Z. 2: 503, t. 100/1 (1966). Type: Zambia, Mbala [Abercorn], Itembe Gorge, *Richards* 12070 (K!, holo., SRGH, iso.)

Shrub or small tree 2.5–7.5 m. tall; branchlets ferruginous pubescent; older ones purplish brown, ridged, lenticellate, ± glabrous. Leaves trifoliolate; petiole 3–7 cm. long, ferruginous pubescent; leaflets discolorous, drying purplish brown above, elliptic, elliptic-oblong or ± obovate, 3.5–10 cm. long, 1.5–4.5 cm. wide, acuminate at the apex, cuneate at the base, shallowly crenate with upwardly directed teeth at tips of crenae, thin, ± adpressed pilose above, rather sparsely pilose to ± densely pubescent beneath with adpressed white hairs and tufts in the axils of the nerves and also along the nerves at secondary nerve junctions; lateral nerves in 8–10 pairs. Inflorescences slender, 5–13 cm. long, unbranched, pubescent; peduncles 4–5 cm. long; cymules 2–3-flowered, subsessile; pedicels 2–3 mm. long. Flowers white, cream or pale green; outer sepals* 1.3 mm. long, 1.1 mm. wide, inner* oblate, 1.5 mm. long, 2–2.5 mm. wide, ± glabrous; petals spathulate, 1.2 mm. long, 0.8 mm. wide, long-ciliate. Filaments 1–1.5 mm. long in male flowers, 0.8 mm. long in female. Ovary hairy; style 1.2 mm. long. Fruit subglobose or ellipsoid, 5–7 mm. long, 4–6 mm. wide, with sparse hairs.

TANZANIA. Iringa, 1 Mar. 1932, *Lynes* I.h.183!
DISTR. **T** 7; Zambia and Malawi
HAB. 'Amongst granite blocks on west ridge'

NOTE. This sheet cited above certainly agrees with several from Zambia named by Exell. A number of specimens of *A. rubifolius* var. *rubifolius* has a similar pattern of hair-tufts.

14. **A. congolanus** *Gilg* in E.J. 24: 294 (1897); Radlk. in Sitz. Akad. Wiss. München 38: 220 (1909) & Baker f. in J.B. 57: 184 (1919); Radlk. in E.P. IV, 165: 527 (1932); T.T.C.L.: 554 (1949); Hauman in F.C.B. 9: 314 (1960); F.Z. 2: 498 (1966). Type: Zaire, M'Towa, 1895, *Descamps* (B†, holo., BM!, fragment, BR, iso.)

Slender several-stemmed erect scrambling or ± scandent shrub or small tree, 1.5–6 m. tall, monoecious (*fide* Hauman); bark grey-brown or purplish black, smooth, conspicuously lenticellate; branchlets ferruginous-velvety-tomentose. Leaves 1-foliolate or 3-foliolate with lateral leaflets very much reduced; petiole 2–4(–7) mm. long, ferruginous-tomentose; petiolules up to 2 mm. long; terminal leaflet elliptic or obovate-elliptic, 5–17.5 cm. long, 3–10.5 cm. wide, acute, subacuminate or obtuse at the apex, cuneate to rounded at the base, serrulate or crenulate-serrulate, velvety-tomentose on both surfaces or eventually pilose or pubescent above; lateral nerves in 8–12 pairs; lateral leaflets ± rounded, 1–3 cm. long and wide. Inflorescences slender, unbranched, 6–20 cm. long, including 2–6 cm. long peduncle; flowers several in subsessile glomerules; pedicels up to 1.5 mm. long, pubescent. Sepals yellow-green; outer ± round, 1.1 mm. long, 1 mm. wide, pubescent; inner 1 mm. long, 0.7 mm. wide pubescent; petals white, spathulate, 1.2 mm. long, 0.5 mm. wide. Filaments 1.5 mm. long; ovary 1.5 mm. long hairy; styles greenish white, 1.5 mm. long. Fruits green becoming orange then bright orange-vermilion, of 1–2 mericarps, each globose, 7 mm. diameter, sparsely pubescent.

TANZANIA. Buha District: Kakombe valley, 6 Feb. 1964, *Pirozynski* 327!; Iringa District: Image, 27 Feb. 1962, *Polhill & Paulo* 1617!; Songea District: 2.5 km. SW. of Kitai, Nakawali R., 7 Mar. 1956, *Milne-Redhead & Taylor* 9049!
DISTR. **T** 4, 6–8; Zaire, Zambia and Malawi
HAB. Grassland with scattered trees or shrubs, lakeside woodland and dry bushland in limestone gorges; 500–1550(–1800 *fide* Goetze) m.

* Exell has them reversed

SYN. *A. appendiculato-serratus* Gilg in E.J. 30: 348 (1901). Type: Tanzania, Njombe District, Livingstone Mts., Yawuanda Mt., *Goetze* 853 (B†, holo.) – Brenan's referral of this to *A. volkensii* in T.T.C.L. is a slip

A. congolanus Gilg var. *monophyllus* Baker f. in J.B. 57: 184 (1919). Type: ?Tanzania, E. coast of Lake Nyasa, *W.P. Johnston* 43 (K!, holo.)

NOTE. The material collected by Milne-Redhead & Taylor shows that unifoliolate and 3-foliolate leaflets occur on the same plant, even the same branch. Exell and Hauman record the species for Kenya but I have seen no correctly named material from there.

15. **A. vestitus** *F.G. Davies* sp. nov., *A. pervillei* Blume valde affinis, sed ramulis junioribus, foliis, petiolis, inflorescentibus fructibusque dense pilosis, pilis setiformibus usque 3 mm. longis differt, etiam *A. congolano* Gilg similis, sed foliolis terminalibus 20–30 × 9–15 cm., rhachidibus inflorescentiarum, petiolis fructibusque longe patente pilosis, cymulis inflorescentiarum regulariter arcte dispositis diversa. Typus: Tanzania, Morogoro District, Manyangu Forest Reserve, *Mgaza* 297 (K!, holo., EA!, K!, iso.)

Small tree 3.6–4 mm. tall with young stems, leaves, petioles, inflorescences, fruits, etc. very densely covered with soft but ± bristle-like hairs up to 3 mm. long, spreading on the stems etc., but ± adpressed on the leaves; older stems grey or purplish brown, densely lenticellate, glabrescent. Leaves discolorous, ± thin, 3-foliolate but with lateral leaflets very much smaller than terminal one or virtually absent; petiole 4–6 cm. long; terminal petiolule 0.4–1.1 cm. long, lateral petiolules 0–1 mm. long; terminal leaflet elliptic or ovate-elliptic, 20–30 cm. long, 9–15 cm. wide, tapering to a narrow acute apex, cuneate at the base, irregularly serrate or crenate-serrate; lateral nerves in 9–15 pairs; lateral leaflets 1.5–4 cm. long, 0.5–2 cm. wide. Inflorescences unbranched, ± 6 cm. long, with closely placed 5–10-flowered sessile cymules, their persistent rhachises ± 1 mm. long, prominently scarred; peduncles up to ± 1 cm. long. Sepals up to 2 mm. long and wide, glabrous. Style 1 mm. long, hairy; stigma-branches 0.8 mm. long. Fruits subglobose, 8 mm. long, 6 mm. wide, spreading hairy.

TANZANIA. Morogoro District: Manyangu Forest Reserve, 1 Aug. 1959, *Mgaza* 297!; Pemba I., Ngezi Forest Reserve, Dec. 1983, *Rodgers, Mwasumbi & Hall* 2772!

DISTR. **T** 6; **P**; not known elsewhere

HAB. Lowland rain-forest (*Odyendea, Antiaris* & *Erythrophleum*) and upland evergreen forest; 10–?1500 m.

NOTE. Clearly closely related to *A. pervillei*.

16. **A grotei*** *F.G. Davies & Verdc.* sp. nov., *A. melliodori* Radlk. valde affinis sed cymulis inflorescentiae distincte pedunculatis, nervis lateralibus longioribus, margine folii haud crenato-serrato sed ± undulatis, domatiis hirsutis deficientibus, ramulis pallidioribus differt. Typus: Tanzania, Lushoto District, E. Usambara Mts., Kwamkoro–Potwe, *Greenway* 2975 (K!, holo., EA, iso.)

Much-branched tree to 9 m. tall; branchlets pale grey-brown, ± pubescent near nodes, glabrescent. Leaves unifoliolate, thin, drying brown, darker above, elliptic to oblong-elliptic, 6.5–12.5 cm. long, 2.5–5.7 cm. wide, narrowly acuminate at the apex, cuneate at the base, undulate or almost entire, glabrous; main lateral nerves ± 6; petiole 0.9–1.5 cm. long; petiolule ± thickened and channelled, 3–5 mm. long. Inflorescences unbranched, 4–7 cm. long, including the 2–2.5 cm. long peduncle; cymules 3–?-flowered with stalks 2–4 mm. long; pedicels 1.5 mm. long; all parts

* A name proposed by Gilg for a gathering from Amani a duplicate of which is at EA.

finely adpressed puberulous. Flowers greenish white; outer sepals ± round, 1.2 mm. long, ± 1 mm. wide, inner similar but wider, all minutely adpressed puberulous. Fruits (see note) ellipsoid, 9.5 mm. long, 5.5 mm. wide, basally slightly ribbed in dry state, glabrous save for very sparse short adpressed hairs. Seeds chestnut, ± ovoid, 5 mm. long.

TANZANIA. Lushoto District: E. Usambara Mts., Kwamkoro–Potwe, 19 June 1932, *Greenway* 2975! & Amani, *Grote*
DISTR. **T** 3; not known elsewhere
HAB. *Ocotea-Cephalosphaera* evergreen rain-forest; 1050 m.

SYN. *A. sp.* sensu T.T.C.L.: 552 (1949)

NOTE. Although there is no adequate reason to reject the seeds as not belonging they are loose in a packet and the portions of plant on the two sheets of 2975 are in bud only and there is no trace of attached fruits; it is strange also that Greenway does not mention fruit colour in his field note. He states 'not very common'; it does not appear to have been seen since 1932. E.G. Baker saw a Grote specimen from Usambara at B.

17. **A. melliodorus** *Radlk.* in Sitz. Akad. Wiss. München 38: 217 (1909) & in E.P. IV, 165: 519 (1932); T.T.C.L.: 552 (1949). Type: Tanzania; E. Usambara Mts., Ngwelo [Nguelo], *Scheffler* 54 (B†, holo., BM!, iso.)

Tree 5–12 m. tall with rough bark; young branchlets dark with pale lenticels, puberulous but soon glabrous; wood yellowish, hard. Leaves unifoliolate, ± thin, elliptic to elliptic-lanceolate, 3.8–14(–18) cm. long, 2–7 cm. wide, rounded or obtuse to mostly acuminate at the apex, cuneate at the base, remotely serrate or crenate-serrate, glabrous except for the conspicuous tufts of hair in the main nerve-axils beneath and often in secondary axils also, not shiny above; lateral nerves in 8–9 pairs, the venation not or scarcely raised above; petiole 1.5–3 cm. long. Inflorescences 6–12 cm. long, simple or with two or more branches; rhachis puberulous; cymules sessile but pedicels 1.5 mm. long. Flowers yellow-green or pale yellow. Sepals rounded, outer ovate, 1 mm. long, 0.8 mm. wide, inner 1.5 mm. long, 1.2 mm. wide. Petals spathulate, 1 mm. long, ciliate. Filaments pubescent at base. Ovary 1 mm. diameter, pubescent; style and stigma together ± 1.5 mm. long. Fruit subglobose, 5.5 mm. diameter, finely pubescent.

TANZANIA. Lushoto District: E. Usambara Mts., Derema [Nderema], 15 July 1956, *Tanner* 2999!; Tanga District: path to Mt. Mlinga from Magrotto, 12 July 1913, *Grote* in *A.H.* 7402!; Morogoro District: W. slopes of Nguru Mts., above Maskati, 17 Mar. 1988, *Bidgood et al.* 531!
DISTR. **T** 3, 6; not known elsewhere
HAB. Rain-forest, montane *Albizia-Podocarpus* forest; 850–1500 m.

NOTE. Vollesen (Opera Bot. 59: 58 (1980)) records a form of this from lowland evergreen forest in **T** 6 (*M.R.C.* 3682) stating that it seems to differ only in being a small shrublet instead of tree; I have not seen it. Radlkofer describes the leaves as 'nitidula' but one of the differences from *A. pervillei* is that the leaves are not shiny above. There are sometimes as Radlkofer points out, minute stipelliform rudiments representing the lateral leaflets.

18. **A. pervillei** *Blume*, Rumphia 3: 123 (1847); Taub. in P.O.A. C: 250 (1895); Radlk. in E. & P. Pf. III, 5: 312 (1895) & in Sitz. Akad. Wiss. München 38: 217 (1909) & in E.P. IV, 165: 517 (1932); T.T.C.L.: 552 (1949); K.T.S.: 505 (1961); Vollesen in Opera. Bot. 59: 58 (1980); S.A. Robertson, Fl. Pl. Seychelles: 44 (1989); Friedmann, Fl. Seychelles Dicot.: 409, t. 118 (1994); Beentje, K.T.S.L.: 413, map & fig. (1994). Type: Madagascar, *Pervillé* 514 (L, holo., B, P, K!, iso.)

Almost invariably a small or medium-sized shrub, but occasionally a small tree or even described as a climber (*Bally* 4691) or prostrate shrub (*Adams* 74), usually 0.9–4(–5) m. tall (but 6–9 m. in Seychelles); bark pale, smooth or rough. Leaves drying discolorous, often bright brown beneath, sometimes somewhat shiny, unifoliolate or trifoliolate with lateral leaflets very much reduced; petiole 0.5–1.5 cm. long; terminal leaflet elliptic or oblong to obovate or obovate-oblanceolate, 3.7–17(–20) cm. long, 1.5–9.6 cm. wide, ± obtuse to acute or acuminate at the apex, cuneate at the base, almost entire to repand-dentate, the nerves ending in hydathodes, or ± coarsely undulate-lobed, rarely distinctly serrate, glabrous save for hairy domatia beneath on some leaves; lateral nerves in 5–7 pairs, often much paler both above and beneath, yellowish, together with the closely reticulate venation, prominent on both surfaces when dry. Inflorescences unbranched, 2–12.5 cm. long, including peduncle 1–4 cm. long (see note); cymules closely placed, 2–7-flowered, the stalks 0–1.2 mm. long; axes glabrous or puberulous; pedicels ± 1 mm. long. Flowers greenish, white or cream; sepals green, the outer ± round, 1.2 mm. long, 1 mm. wide; inner ± oblate, 1–1.5 mm. long, 1.2–2 mm. wide, finely adpressed-pubescent or glabrous; petals spathulate, 1.5 mm. long, 0.8 mm. wide, mostly densely white hairy or ciliate to ± glabrous; disk glabrous. Filaments, ovary and style densely hairy. Fruits bright red, globose or ± obovoid, 7–9 mm. long, 6–7 mm. wide, finely pubescent.

forma **pervillei**

Leaves unifoliolate, with no trace of lateral leaflets. Fig. 24.

KENYA. Kwale District: Shimba Hills National Reserve, 28 Jan. 1983, *S.A. Robertson* 3502!; Kilifi District: Arabuko, May 1929, *R.M. Graham* in *F.D.* 2126!; Lamu District: Utwani Forest, Dec. 1956, *Rawlins* 275!

TANZANIA. Pangani District: Kumbamtoni, Sept. 1955, *Semsei* 2272!; Morogoro District: Kimboza Forest Reserve, July 1952, *Semsei* 782!; Zanzibar, Mazizini [Massazine], 30 May 1959, *Faulkner* 2279!; Pemba, Kengeja, 5 Mar. 1952, *R.O. Williams* 140!

DISTR. **K** 7; **T** 3, 6; **Z**; **P**; Madagascar, Comoro Is. and Seychelles

HAB. Fringing forest, forest clumps, ground-water forest, coastal *Brachystegia* and coral and other limestone, cliff tops etc.; 0–550 m. (see note)

SYN. [*Schmidelia monophylla* sensu Baker in F.T.A. 1: 424 (1868), *non* (E. Mey.) C. Presl]
[*Allophylus monophyllus* sensu Taub. in P.O.A. C: 250 (1895), *non* (E. Mey.) Radlk.]
A. cobbe (L.) Räusch. ('*pervillei*'); Capuron in Mém. Mus. Nat. Hist. Nat., Paris, 19: 63, t. 13/1–5 (1969)

NOTE. A variant occurs at 750–900 m., *D.W. Thomas* 3635 (Tanzania, Iringa District, Mwanihana Forest Reserve, above Sanje, 30 Aug. 1984) with leaves to 20 × 9.5 cm. and inflorescences to 18 cm. long. A few specimens have some inflorescences densely tomentose, e.g. *R.M. Graham* in *F.D.* 2126, but I suspect this is due to disease rather than a natural variation. Occasionally the flowers are galled resulting in spiky galls covered with dense long processes ± 3 mm. long; these resemble miniature versions of the fruits often found in some genera of Flacourtiaceae.

forma **trifoliolatus** *Radlk.* in E.P. IV, 165: 517 (1932). Types: Madagascar, Nosy Be, *Boivin* 47 (P, syn.) & Nosy Be, Sakatia I., *Hildebrandt* 3363 (B†, syn.) & Seychelles, *Wright* (K!, syn.)

Leaves trifoliolate with very small lateral leaflets 1.5–2 cm. long, 0.8–1.3 cm. wide.

KENYA. Kwale District: 13 km. WSW. of Gazi, Buda Mafisini Forest, 22 Aug. 1953, *Drummond & Hemsley* 3952!; Kilifi District: Vipingo Estate, 28 Dec. 1945, *Bally* 4685!; Lamu District: Mkune Swamp, 1 Mar. 1957, *Rawlins* 349!

TANZANIA. Morogoro District: 6.5 km. N. of Turiani, Lusunguru Forest Reserve, near Mtibwa sawmill, 31 Mar. 1953, *Drummond & Hemsley* 1932!; Uzaramo District: 34 km. S. of Dar es Salaam on Kilwa road, 17 Feb. 1970, *B.J. Harris & Tadros* 4181!; Rufiji District: Mafia I., 28 Mar. 1933, *Wallace* 781!; Zanzibar, Chwaka, 7 Sept. 1959, *Faulkner* 2353!

DISTR. **K** 7; **T** 6; **Z**; Madagascar and Seychelles

HAB. Coastal forest bushland and thicket; ground-water forest; 0–400 m.

FIG. 24. *ALLOPHYLUS PERVILLEI* forma *PERVILLEI* — **1**, flowering and fruiting branch, × $^{2}/_{3}$; **2**, male flower, × 10; **3**, male flower with four sepals removed, × 10; **4**, petal from male flower, × 12; **5**, female flower, × 10; **6**, petal from female flower, × 10. 1, from *Faulkner* 2279; 2–6, from *S.A. Robertson* 3502. Drawn by Pat Halliday.

IMPERFECTLY KNOWN TAXA

A. bussei *Gilg* in V.E. 3(2): 270 (1921), *nom. nud.*; T.T.C.L.: 555 (1949)

This has not been described and is not mentioned in Radlkofer's E.P. revision. The specimen came from the Rondo [Muera] Plateau, Lindi District, but has not been seen. E.G. Baker on his sketch (BM) reports that the inflorescence is distinctly branched with branches sometimes 8–9 cm. long and the leaflets very distinctly acuminate. It is almost certainly *A. chirindensis* Baker f.

A. holstii *Gilg* in V.E. 3(2): 270 (1921), *nom. subnud.*; T.T.C.L.: 555 (1949)

This has not been described and is not mentioned in Radlkofer's E.P. revision. The specimen came from Dar es Salaam but has not been seen. E.G. Baker saw a sheet labelled *A. holtzii* at B. Holtz collected around Dar es Salaam and I suspect *holstii* is a slip. It is a constantly recurring error. Baker's rough sketch suggests *A. congolanus.*

A. rutete *Gilg* in Z.A.E.: 476 (1912); Radlk. in E.P. IV, 165: 547 (1932); T.T.C.L.: 555 (1949). Type: Uganda, Ankole District, N. of the Kagera R., between Nyavatura and Kesimbili, *Mildbraed* 318 (B†, holo.)

I think this must be a glabrous form of *A. fulvotomentosus* Gilg to which Gilg says it is 'nahe verwandt'; the phrase 'inflorescentiis numerosissimis densissimis ornatus' certainly agrees but 'petiolo glabro' and leaflets "utrinque glaberrimis' do not fit material I have seen. E.G. Baker's very rough sketch made at Berlin confirms the leaves as being glabrous and bears a note 'near *A. fulvotomentosus*'.

24. CARDIOSPERMUM

L., Sp. Pl.: 366 (1753) & Gen. Pl., ed. 5: 171 (1754); Radlk. in E.P. IV, 165: 370 (1932)

Monoecious annual or sometimes perennial herbaceous or woody climbers, with ribbed stems, some species, including all those in Africa, climbing by means of tendrils. Leaves variously compound on a trifoliate basis: ternate with deeply 3-lobed leaflets, biternate or (not in Africa) many-times ternately compound or pinnate. Inflorescences axillary, each a branched thyrse; commonly with paired tendrils at the top of the peduncle, which may itself act as a tendril. Flowers slightly irregular; sepals (in Africa) 4, unequal; petals 4, each with a scale-like appendage directed away from the petal, together forming an additional perianth whorl; disk unilateral, extrastaminal, consisting of two variously-shaped glands. Male flowers with 8 dorsifixed stamens and minute pistillode. Female flower with 8 staminodes; ovary 3-celled, style trifid. Fruit an inflated capsule with 3, 1-seeded locules. Seeds black, globose, the hilum conspicuously white, covered when fresh by an aril.

Three species of this predominantly tropical American genus are native to Africa and a fourth, *C. integerrimum* Radlk. has been introduced into Cameroon.

Radlkofer includes species 1 in sect. *Ceratadenia* Radlk., with the disk elements larger and elongate, and species 2 and 3 in 'sect. *Brachyadenia* Radlk.', with them shorter and subglobose. Since this contains the type species of the genus it is illegitimate and should be called sect. *Cardiospermum.*

1. Petals small, 2–3 mm. long; fruit broadest at apex (i.e. distally); leaves glabrescent or sparsely pubescent · · · · 2. *C. halicacabum*
   Petals larger, (3–)4–9 mm. long; fruit broadest near middle; leaves glabrescent to densely velvety · 2

2. Petals larger, 8–9 mm. long; fruit large, 4.5–7 cm. long, ± glabrous; leaves glabrescent or sparsely pubescent (in Flora area), often with tufts of hairs in axils of lateral nerves beneath · 1. *C. grandiflorum*
Petals smaller, (3–)4–6 mm. long; fruit smaller, 2–4 cm. long, pubescent; leaves usually densely velvety pubescent 3. *C. corindum*

1. **C. grandiflorum** *Sw.*, Prodr. 64 (1788); P.O.A. C: 249 (1895); Radlk. in E.P. IV, 165: 372 (1932); F.P.N.A. 1: 515 (1948); F.P.S. 2: 338, fig. 124 (1952); C.F.A. 2: 74 (1954); F.W.T.A., ed. 2, 1: 711, fig. 197 (1958); Hauman in F.C.B. 9: 285 (1960); Exell in F.Z. 2: 511, t. 101B (1966); F.P.U.: 68, fig. 24 (1972); Fouilloy & Hallé in Fl. Cameroun, 16: 17, t. 1/10–17 (1973) & in Fl. Gabon 23: 17, t. 1/10–17 (1973); U.K.W.F., ed. 2: 164 (1994). Type: Jamaica, *Swartz* (S, holo.)

Herbaceous or slightly woody climber, 0.9–8 m. long; stems deeply ridged, with short curled pubescence or with long spreading brown or golden brown hairs 3–5 mm. long. Leaves biternate or less often ternate, up to 18 cm. long, with deeply 3-lobed leaflets; petiole up to 6 cm. long, grooved, glabrous or pubescent; first order petiolules 1–3 cm. long, the central one longer or shorter than the laterals, those of ultimate leaflets 0–1 cm. long; leaflets ovate-rhombic, 1.5–9 cm. long, 1.2–4.3 cm. wide, acuminate at the apex, tapering at the base, coarsely crenate, ± glabrous or with tufts of hairs in the nerve axils beneath (can be velvety in South America); lateral nerves 5–12, weak. Inflorescences spreading horizontally, paniculate, with a pair of tendrils at the apex of the peduncle; peduncle 5–10 cm. long; axis 5–10 cm. long; pedicels sometimes orange, 1–5 mm. long. Flowers sweetly scented, 6–11 mm. long; sepals 4, greenish white sometimes crimson flushed outside, unequal, the shorter 3–3.5 mm. long, 2.5 mm. wide, the longer petaloid, 6–7 mm. long, 5 mm. wide; petals white, ovate, 8–9 mm. long, scale orange-yellow, contiguous in the centre of the flower to form a cup or corona enclosing the stamens or ovary; disk extrastaminal, formed of 2 adjacent club-shaped parts 5 mm. long. Stamens white. Stigma linear, 2–5 mm. long. Fruits green, ellipsoid, 4.5–7 cm. long, 2–3 cm. wide, membranous, inflated, strongly venose; septa 3, white or pale brown, papery, 5 cm. long, 3 cm. wide, remaining after dehiscence and each bearing a seed, forming a wind sailing wing. Seed black, globose, 7 mm. diameter, glabrous, with a white rounded-ovate hilum 1.5–2 mm. long. Fig. 25/5–7.

UGANDA. Ankole Forest: Kalinzu Forest, 21 May 1961, *Symes* 731!; Mbale District: 20 km. W. of Busia on Jinja road, W. Bugwe Forest Reserve, 6 Sept. 1950, *Dawkins* 632!; Mengo District: Entebbe, lake shore, Oct. 1931, *Eggeling* 27!

KENYA. W. Suk District: 80 km. N. of Kapenguria, Weiwei R. valley, Sigor, 13 Nov. 1953, *Bogdan* 3834! & Cherangani Hills, edge of R. Morun, near Sebit, 31 July 1969, *Mabberley & McCall* 99!; N. Kavirondo District: Kakamega Forest, Yala R. area, 10 Dec. 1956, *Verdcourt* 1701!

TANZANIA. Mwanza District: Geita, Juma Is., 25 Jan. 1953, *Tanner* 1182!; Kigoma District: Gombe Stream Reserve, Mkenke stream, 15 Apr. 1964, *Pirozynski* 700!; Buha District: 16 km. N. of Kigoma, Kakombe, 7 July 1959, *Newbould & Harley* 4267!

DISTR. **U** 2–4, **K** 2, 3, 5; **T** 1, 4; also cultivated in **K** 4; widespread in tropical Africa to Namibia and South Africa, central and tropical America, also introduced into Australia, Sri Lanka, Malaya and southern Europe

HAB. Riverine forest and thicket, grassland patches within forest, roadsides, usually in marginal situations; 750–1500 m.*

SYN. *C. hirsutum* Willd., Sp. Pl., ed. 4, 2: 467 (1799). Type: Ghana/Benin [Guinea], *Isert* in *Herb. Willd.* 7732 (B-W, holo.)
C. elegans Kunth, Nov. Gen. Sp. 5: 99, t. 439 (1821). Type: Peru, *Humboldt & Bonpland* (P, holo.)
C. barbicaule Baker in F.T.A. 1: 418 (1868). Type: Cameroon, Mt. Cameroon, *Mann* 1299 (K!, holo.)

* Davies gives 1–2000 m., but this must include records from outside the Flora area. [B.V.]

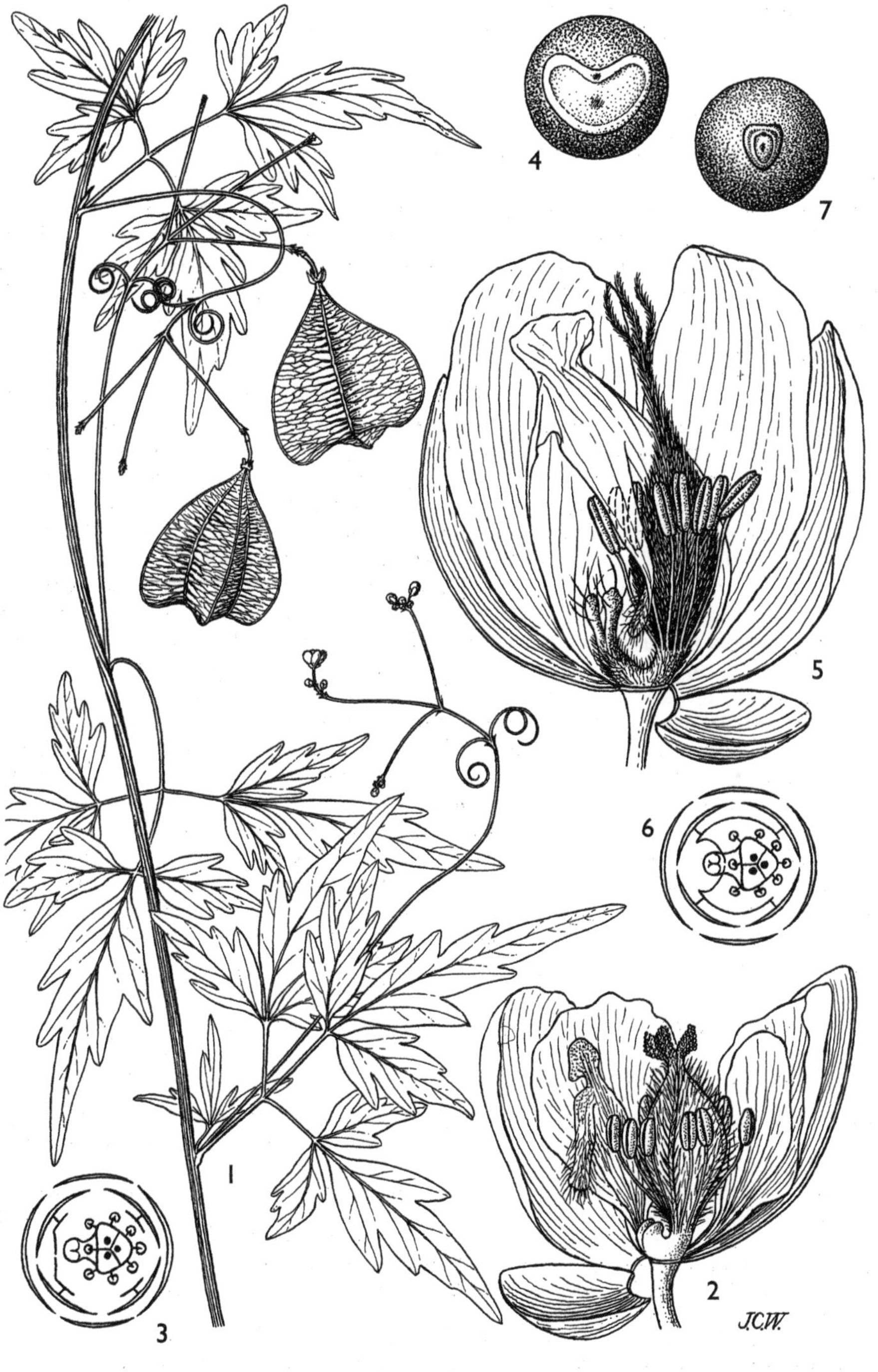

FIG. 25. *CARDIOSPERMUM HALICACABUM* — **1**, branch with flowers and fruit, × 1; **2**, flower with two petals removed and sepals displayed, × 15; **3**, diagram of flower; **4**, seed, × 6. *CARDIOSPERMUM GRANDIFLORUM* — **5**, flower with two petals removed and sepals displayed, × 9; **6**, diagram of flower; **7**, seed, × 4.5. 1–4, from *Chase* 2311; 5–7, from *Banda* 177. Reproduced from F.Z. 2, t. 101 (1966).

C. grandiflorum Sw. forma *hirsutum* (Willd.) Radlk. in Sitz. Akad. Wiss. München 8: 260 (1878) & in E.P. IV, 165: 375 (1932)
C. grandiflorum Sw. forma *elegans* (Kunth) Radlk. in Sitz. Akad. Wiss. München 8: 260 (1878) & in E.P. IV, 165: 375 (1932)

2. **C. halicacabum** *L.*, Sp. Pl.: 366 (1753); F.T.A. 1: 417 (1868); P.O.A. C: 249 (1895); Radlk. in E.P. IV, 165: 379, fig. 8A–C (1932); F.P.N.A. 1: 515 (1948); U.O.P.Z.: 171 (1949); F.P.S. 2: 339, fig. 125 (1952); C.F.A. 2: 75 (1954); F.W.T.A., ed. 2, 1: 711 (1958); Hauman in F.C.B. 9: 286 (1960); Exell in F.Z. 2: 509, t. 101A (1966); Fouilloy & Hallé in Fl. Cameroun 16: 18, t. 1/18–22 (1973); & in Fl. Gabon 23: 18, t. 1/18–22 (1973); Vollesen in Opera Bot. 59: 58 (1980); Troupin, Fl. Rwanda 2: 303 (1983); Vollesen in Fl. Ethiopia 3: 498 (1990); Leenh. in Fl. Males., ser. 1, 11: 484, fig. 12 (1994); U.K.W.F., ed. 2: 164, t. 60 (1994). Type: plant grown in Clifford's garden in Holland, *Hort. Cliff.* 151 (BM!, lecto.)

Annual herbaceous or slightly woody climber, 0.3–3 mm. long; stems deeply grooved, straw-coloured with grooves green. Leaves pale green, stipulate, thin, biternate, triangular in outline, (3–)5–10 cm. long, 5–12 cm. wide, composed of 3 ± equal divisions each of a petiolule and 3 leaflets or deeply trifid leaflet; petiole 0.4–6 cm. long, pubescent; first order petiolules 0.5–3 cm. long, subequal or often the lateral ones somewhat longer than median one; leaflets sessile or with short secondary petiolules, sparsely pubescent when young, later glabrescent except for main nerves, the terminal one ovate, 1.5–6.5 cm. long, 0.8–2.5 cm. wide, acute to acuminate or rarely rounded at the apex, apiculate, attenuate to rounded at the base, serrate to deeply dentate or pinnatipartite, rarely ± entire. Inflorescences corymbose, the rhachis bearing 3–4, 2–5-flowered raceme-like thyrses; peduncle below tendrils 1.5–10.5 cm. long, the part above 0.2–1.2 cm. long; secondary peduncles 0.3–1.7 cm. long; pedicels (0–)1–2 mm. long, lengthening to 5 mm. in fruit. Flowers said not to be scented; sepals yellow-green, unequal, the outer ± round, 1 mm. diameter, the inner larger and ovate, 2–3 mm. long, 1–2 mm. wide; petals white, cream, greenish or yellow, elliptic to obovate, 2–3 mm. long; appendages yellow at tip, hairy; disk of 2 curved elements 1 mm. long. Stamens 8; filaments ± 2 mm. long, pilose. Ovary ± 2 mm. diameter; style 0.5 mm. long, with yellow stigma equalling it. Fruit yellow-green turning greenish brown to reddish or dark brown, obtetrahedral, 1–3.5 cm. long, 1.8–4 cm. wide, dehiscing into 3 segments, at first densely puberulous or pubescent. Seeds black, globose, 5 mm. diameter, with a reniform or crescentic hilar area, 3–4 mm. wide. Fig. 25/1–4.

var. **halicacabum**; Exell in F.Z. 2: 509 (1966); Vollesen in Fl. Ethiopia 3: 498, fig. 125.6 (1990)

Capsules larger, 3–3.5 cm. long, 3.2–4 cm. wide.

UGANDA. W. Nile District: 1.5 km. SE. of Maracha Rest Camp, 26 July 1953, *Chancellor* 52!; Mbale District: W. of Elgon, Aug. 1959, *Tweedie* 1887!; Mengo District: Port Bell, 29 June 1951, *Norman* 47!

KENYA. Northern Frontier Province: Marsabit, July 1942, *J. Bally* in *Bally* 1869!; Machakos District: Nairobi–Mombasa road, Athi R., 30 May 1958, *Verdcourt & Napper* 2179!; Kilifi District: Kilifi–Kaloleni road, Mitangoni, 22 June 1969, *Adams* 21!

TANZANIA. Musoma District: Ikizu, 13 Apr. 1959, *Tanner* 4138!; Singida District: Lake Kitangiri flats, Tulia, 29 Apr. 1962, *Polhill & Paulo* 2249!; Morogoro District: Mgeta Bank, Oct. 1860, *Grant*!

DISTR. **U** 1, 2*, 3, 4; **K** 1, 2, ?3, 4, 5, 7; **T** 1–8; pantropical

HAB. Grassland, open bushland, *Acacia* etc. woodland, edges of flood plains, cultivations; 50–1700 m.

* *Doggett* from **U** 2, Toro has been named var. *microcarpum*, but the fruits are not mature and its position dubious. Also recorded from **K** 6 in U.K.W.F., but varieties not separated there.

var. **microcarpum** (*Kunth*) *Blume*, Rumphia 3: 185 (1847); Radlk. in E.P. IV, 165: 391 (1932); Exell in F.Z. 2: 509 (1966); Fouilloy & Hallé in Fl. Cameroun 16: 19 (1973) & in Fl. Gabon 23: 19 (1973); Vollesen in Fl. Ethiopia 3: 498 (1980); F. Friedmann in Fl. Masc. 76: 5, t. 1/4–8 (1997). Type: Venezuela, near S. Fernando de Atabapo [Missiones del Orinoco], *Humboldt & Bonpland* (P, holo.)

Capsule smaller, 1–1.5 cm. long, 1.8–2.5 cm. wide (typically in South America, 1 cm. long, 1.8–2 cm. wide).

KENYA. S. Kavirondo District: Mbita Point Field Station, 30 Nov. 1981, *Gachathi & Opon* 102/81!; Masai District: Ol Lorgasailie, 25 Aug. 1963, *Mackinnon* 26!; Coastal Province: Siani, 17 Mar. 1902, *Kassner* 339!

TANZANIA. Rufiji District: Mafia I., Ras Mbisi, 7 Aug. 1932, *Schlieben* 2645!; Iringa District: 5 km. on Msembe [Msembi]–Mbagi track, 5 km., 31 Mar. 1970, *Greenway & Kanuri* 14234!; Zanzibar, Bububu [Bu-Bu-Bu], 27 Jan. 1930, *Vaughan* 1151!; Pemba, Chake Chake, 18 Sept. 1929, *Vaughan* 659!

DISTR. **K** 5–7; **T** 3, 5, 6 (Mafia), 7; **Z**; **P**; pantropical

HAB. Coastal grassland on coral rock, also coconut plantations, inland in maize plots, lake shores and bushland sometimes on volcanic soil; 0–1400 m.

SYN. *C. microcarpum* Kunth, Nov. Gen. Sp. 5: 104 (1821); F.T.A. 1: 418 (1868); F.P.S. 2: 339 (1952); C.F.A. 2: 75 (1954)

C. truncatum A. Rich., Tent. Fl. Abyss. 1: 101 (1847). Type: Ethiopia, Chiré [Shire] & Taccaze valley (P, syn.)

NOTE. A well-marked variety in many areas but intermediates do occur in East Africa.

3. **C. corindum** *L.*, Sp. Pl., ed. 2: 526 (1762); Radlk. in E.P. IV, 165: 397 (1932); F.P.S. 2: 339 (1952); C.F.A. 2: 75 (1954); Exell in F.Z. 2: 509 (1966); Fouilloy & Hallé in Fl. Cameroun 16: 19 (1973) & in Fl. Gabon 23: 19 (1973); Vollesen in Fl. Ethiopia 3: 497 (1990); U.K.W.F., ed. 2: 164 (1994). Type: Corindum villosum, foliis quinatoternatis incisis obtusis, petiolis brevibus, Mill., Gard. Dict., ed. 7 (1759) (holo.!)*

Annual herbaceous to slightly woody climber, 0.3–4.5 m. tall; stems deeply grooved, densely shortly pubescent. Leaves stipulate, biternate or pinnate, triangular or ovate in outline, up to 9 cm. long, 10 cm. wide, mostly densely velvety pubescent, in biternate examples composed of 3 main divisions, the median larger than the laterals; petiole 0.3–2(–3) cm. long; first order petiolules 0.1–3 cm. long, second order petiolules 0–1 cm. long; terminal leaflet ovate, 1–5.2 cm. long, 0.5–4.5 cm. wide, acute to acuminate and apiculate at the apex, attenuate-decurrent to truncate or cordate at the base, crenate to serrate or deeply lobed. Inflorescences corymbose to paniculate, the rhachis with 3–5, few-flowered corymbiform racemoid thyrses, grey-pubescent, or a paniculate thyrse 0.4–1.5 cm. long with 6–12 racemoid thyrses 0.2–1.1 cm. long; peduncle below tendrils (2–)3–12 cm. long, above 0.2–1.3 cm. long; secondary peduncles 0.2–1.4 cm. long; pedicels 2–4 mm. long. Sepals unequal; outer broadly elliptic to ± round, 1.5–2 mm. long, 1.5 mm. wide, inner elliptic, 3–5.5 mm. long, 2–3 mm. wide; petals white or cream, elliptic, (3–)4–6 mm. long, 2–3.5 mm. wide; appendage orange. Stamens 8, with filaments 3–4 mm. long, pilose. Ovary 2–3 mm. long; style 0.5–1 mm. long; stigmas ± 0.5 mm. long. Fruit green to orange or rusty red, ovoid or usually ellipsoid, (2–)2.5–3.5(–4) cm. long, (1.5–)2–3.3 cm. wide, finely velvety when young; septa elliptic, longer than wide. Seeds black, 4.5–5 mm. diameter; hilum white, reniform or cordate, 3 mm. long, 3.5 mm. wide.

UGANDA. Karamoja District: Amudat, 24 July 1954, *J. Wilson* 117!

* Based on a specimen from Jamaica collected by Houston (BM!) but Miller does not actually mention him.

KENYA. Northern Frontier Province: Ndoto Mts., Ndigri–Alori, 31 Dec. 1958, *Newbould* 3399!; Kitui District: Mutomo, Voo location, near Ndui Rock Dam, 30 Aug. 1968, *Kimani* 102!; Teita District: Tsavo National Park East, Voi Park H.Q. hill, SW. outlier, 23 Dec. 1966, *Greenway & Kanuri* 12848!

TANZANIA. Lushoto District: Amani, 18 Oct. 1910, *Braun* 3117

DISTR. **U** 1, **K** 1, 2, 3 (*fide* U.K.W.F.), 4, 6 (*fide* U.K.W.F.) 7; **T** 3 (*fide* EA); Sudan, Ethiopia and Somalia to South Africa, pantropical

HAB. *Acacia-Commiphora* etc. bushland and woodland, reaching to tops of *Acacias*, sometimes riverine, also scrub on rocky hills; 500–1500 m.

SYN. *C. canescens* Wall, Pl. As. Rar. 1: 14, t. 14 (1829); F.T.A. 1: 418 (1868); P.O.A. C: 249 (1895). Type: Wall., Pl. As. Rar., t. 14 (syn.!)*

C. clematideum A. Rich., Tent. Fl. Abyss.: 100 (1847); Schweinf., Beitr. Fl. Aethiop.: 263 (1867). Type: Ethiopia, Taccazé valley, *Quartin Dillon* (P, holo.)

C. oblongum A. Rich., Tent. Fl. Abyss.: 101 (1847). Type: Ethiopia, Taccazé valley, Djeladjeranne, *Quartin Dillon* (P, holo.)

C. corindum L. forma *canescens* (Wall.) Radlk. in Martius, Fl. Bras. 13(3): 448 (1897) & in E.P. IV, 165: 401 (1932)

C. corindum L. forma *clematideum* (A. Rich.) Radlk. in Martius, Fl. Bras. 13(3): 448 (1897) & in E.P. IV, 165: 402 (1932)

NOTE. Radlkofer maintains 14 forms in his Pflanzenreich account but it appears that all these cannot really be upheld. M. Ferrucci has annotated East African material as forma *canescens*.

25. **PAULLINIA**

L., Sp. Pl.: 365 (1753) & Gen. Pl., ed. 5: 170 (1754); Radlk., Monogr. Paullinia: 67–381(1895) & in E.P. IV, 165: 219 (1931)

Monoecious slender woody climbers, often with coiled tendrils at the base of the inflorescence. Leaves often stipulate, petiolate, variously compound, the petiole and rhachis commonly winged. Inflorescence an axillary (or cauliflorous in America) thyrse, simple, branched or clustered. Flowers slightly zygomorphic; sepals 5; petals 4 with scale prominent; disk reduced to little more than a unilateral pair of glands outside the stamens. Stamens and staminodes 8; anthers glabrous; filaments pilose, connate at the base. Ovary 3-locular with 1 ovule per locule; style trifid. Fruit a 3-valved obovoid capsule, often winged (not in Africa), sometimes spiny, dehiscing into 3 valves. Seeds borne on the central axis, black, arillate.

A large genus of ± 200 species in tropical and subtropical America, including one species also throughout tropical Africa.

P. pinnata *L.*, Sp. Pl.: 366 (1753); F.T.A. 1: 419 (1868); P.O.A. C: 249 (1895); Radlk., Monogr. Paullinia: 135 (1895) & in E.P. IV, 165: 247 (1931); F.P.N.A. 1: 514 (1948); U.O.P.Z.: 401, fig. on 403 (1949); T.T.C.L.: 560 (1949); F.P.S. 2: 343, fig. 127 (1952); C.F.A.: 73 (1954); F.W.T.A., ed. 2, 1: 710, fig. 196 (1958); Hauman in F.C.B. 9: 283 (1960); F.F.N.R.: 225 (1962); Haerdi in Acta Trop., suppl. 8: 124 (1964); Exell in F.Z. 2: 512, t. 102 (1966); Fouilloy & Hallé in Fl. Cameroun 16: 13, t. 1/1–9 (1973) & in Fl. Gabon 23: 13, t. 1/1–9 (1973); F.G. Davies in K.B. 32: 429–431 (1978); Vollesen in

* Two specimens at K from Burma *Wallich* 8032 "road to petroleum wells (Irawaddi) 1826" are presumably syntypes; Wallich mentions banks Irrawadi, Pagemew, Prome, Segaen, Ava, also Martabania.

Opera Bot. 59: 58 (1980); Troupin, Fl. Rwanda 2: 310, fig. 96/2 (1983); Vollesen in Fl. Ethiopia 3: 493, fig. 125.5 (1990); Beentje, K.T.S.L.: 423, fig., map on 422 (1994); U.K.W.F., ed. 2: 164 (1994). Lectotype, chosen by F.G. Davies: Plumier, Descr. Pl. Amer.: 76, t. 91 (1693) (lecto.!)

Shrubby climber 2.5–8(–?15) m. high, with a deep brown sharply ribbed stem up to 3.8 cm. wide at base. Leaves imparipinnate, 5-foliolate, (5–)15–25 cm. long, 8–20 cm. wide; petiole up to 10 cm. long, it and the rhachis winged, the wing 0.5–1.3 cm. wide; leaflets sessile or shortly stalked, ovate, elliptic or rhomboid, the terminal one 7–15 cm. long, 3.5–8 cm. wide, the laterals 3.5–10 cm. long, 1.7–5 cm. wide, acuminate, acute or obtuse at the apex, cordate, cuneate or narrowed at the base, subentire (rarely in our area), distantly dentate, serrate or shallowly lobed, subcoriaceous, glabrous, shining, with hairs present beneath on the venation and in tufts in the nerve axils; lateral nerves strong. Inflorescence a racemoid thyrse axillary in each of the upper leaves, 5–10 cm. long; peduncle 1–3 × the length of the fertile portion, with apical pair of tightly coiled tendrils, or occasionally inflorescence sessile; inflorescence-axes brown-puberulent; pedicels 0–3 mm. long; flowers (3–)5–9 in cymose clusters, the terminal one often female and the remainder male. Flowers yellowish, creamy white or pink, scented, 5–6 mm. long; sepals ovate, unequal, outermost 1.5 mm. long, innermost 4 mm. long, pubescent on their outer surfaces; petals white, tongue-shaped, subequal, 5 mm. long, scale asymmetrically crested, ciliate, $^3/_4$ of the length of the petal, yellow-tipped; disk to one side of the androgynophore, ciliate, consisting mainly of 2 large, plus 2 minute, ear-shaped glands. Stamens 8, 5 long and 3 short; anthers white; staminodes 8. Ovary pubescent; style trifid; rudimentary ovary much reduced. Fruit green, turning red or rose-red, obovoid or obovoid-oblong, 2.4–3.6 cm. long, 1.1–1.4 cm. wide, rounded or emarginate at the apex, mucronate, strongly narrowed at the base, 3-grooved, at first pubescent but later glabrous and finely rugulose, ultimately dehiscent. Seeds black, oblong-ellipsoid, 1–1.1 cm. long, 7–8 mm. wide; aril white. Fig. 26.

UGANDA. Kigezi District: Kanunga, Mar. 1946, *Purseglove* 1983!; Ankole District: Ruizi R., 25 Nov. 1950, *Jarrett* 166!; Mengo District: 17 km. on Kampala–Bombo road, Oct. 1930, *Snowden* 1764!

KENYA. Nandi District: Kaimosi, SW. of Yala R. bridge, 18 Apr. 1965, *Gillett* 16699!; S. Kavirondo District: Migori R., 19 Dec. 1957, *Jarrett* SN/4!; Kwale District: Shimba Hills, Shimba Forest, 15 Mar. 1968, *Magogo & Glover* 284!

TANZANIA. Ngara District: Bushubi, Muganza, 7 Sept. 1960, *Tanner* 5145!; Lushoto District: near Derema, 12 Mar. 1950, *Verdcourt* 107!; Kilosa District: Vigude, Kidodi, Nov. 1952, *Semsei* 1042!; Zanzibar, Chwaka, km. 26, 25 Apr. 1960, *Faulkner* 2534!

DISTR. **U** 2–4; **K** 3, 4 (*fide* U.K.W.F.), 5, 6 (*fide* Davies), 7; **T** 1–4, 6–8; **Z**; **P**; throughout tropical Africa, Madagascar, tropical and subtropical America

HAB. Forest margins, gallery forest, moist thicket and scrub; 0–1600 m.

NOTE. The tendrils are sometimes described as axillary and at base of inflorescence – certainly at the apices of shoots some tendrils are found in the axils and it is not clear if all the 'peduncles' bearing apical tendrils but no inflorescences will eventually bear inflorescences. The peduncles may in fact be modified branches. The stems are used for making ropes and for cane and basket work, e.g. fish traps. The plant has been used as a fish poison.

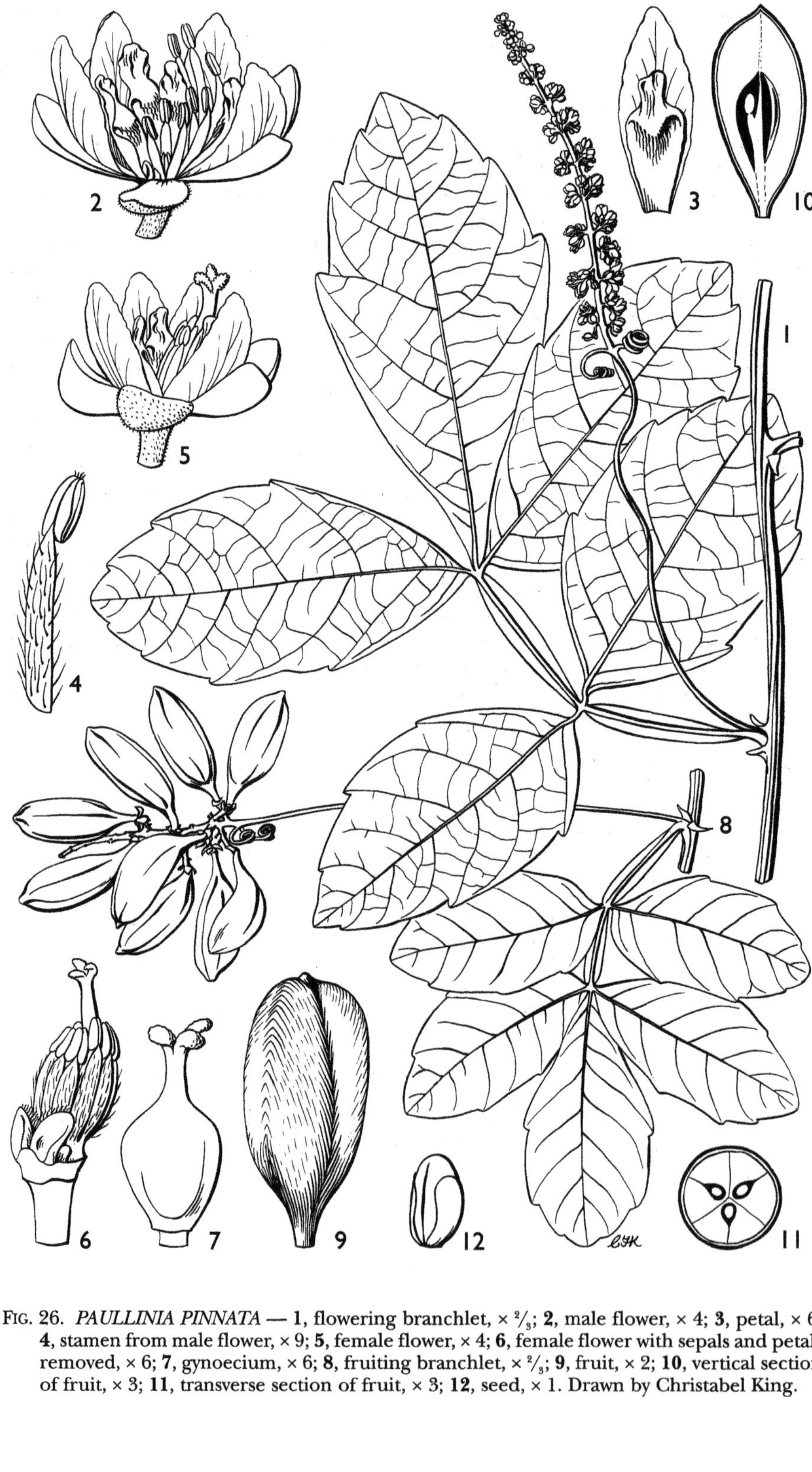

FIG. 26. *PAULLINIA PINNATA* — **1**, flowering branchlet, × $^{2}/_{3}$; **2**, male flower, × 4; **3**, petal, × 6; **4**, stamen from male flower, × 9; **5**, female flower, × 4; **6**, female flower with sepals and petals removed, × 6; **7**, gynoecium, × 6; **8**, fruiting branchlet, × $^{2}/_{3}$; **9**, fruit, × 2; **10**, vertical section of fruit, × 3; **11**, transverse section of fruit, × 3; **12**, seed, × 1. Drawn by Christabel King.

INDEX TO SAPINDACEAE

New names validated in this part

GEOGRAPHICAL DIVISIONS OF THE FLORA

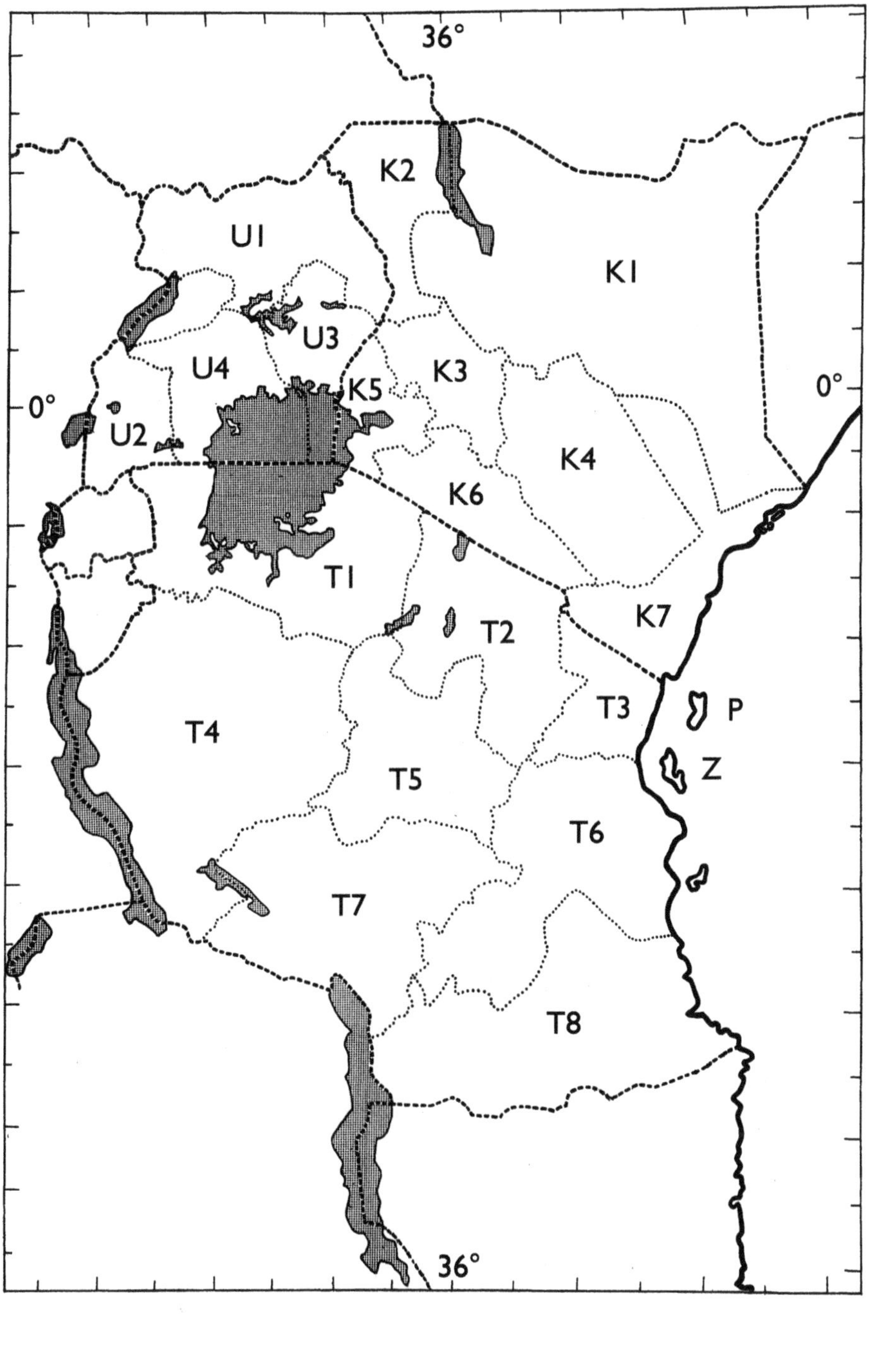